Interview Questions in Business Analytics

Bhasker Gupta

Apress®

Interview Questions in Business Analytics

Bhasker Gupta
Bangalore, Karnataka, India

ISBN-13 (pbk): 978-1-4842-0600-3   ISBN-13 (electronic): 978-1-4842-0599-0
DOI 10.1007/978-1-4842-0599-0

Library of Congress Control Number: 2016948865

Managing Director: Welmoed Spahr
Lead Editor: Celestin Suresh John
Technical Reviewer: Manoj R. Patil
Editorial Board: Steve Anglin, Pramila Balan, Laura Berendson, Aaron Black, Louise Corrigan,
    Jonathan Gennick, Robert Hutchinson, Celestin Suresh John, Nikhil Karkal,
    James Markham, Susan McDermott, Matthew Moodie, Natalie Pao, Gwenan Spearing
Coordinating Editor: Prachi Mehta
Copy Editor: Michael G. Laraque
Compositor: SPi Global
Indexer: SPi Global
Artist: SPi Global

Distributed to the book trade worldwide by Springer Science+Business Media New York, 233 Spring Street, 6th Floor, New York, NY 10013. Phone 1-800-SPRINGER, fax (201) 348-4505, e-mail orders-ny@springer-sbm.com, or visit www.springeronline.com. Apress Media, LLC is a California LLC and the sole member (owner) is Springer Science+Business Media Finance Inc (SSBM Finance Inc). SSBM Finance Inc is a **Delaware** corporation.

For information on translations, please e-mail rights@apress.com, or visit www.apress.com.

Apress and friends of ED books may be purchased in bulk for academic, corporate, or promotional use. eBook versions and licenses are also available for most titles. For more information, reference our Special Bulk Sales–eBook Licensing web page at www.apress.com/bulk-sales.

Any source code or other supplementary materials referenced by the author in this text are available to readers at www.apress.com. For detailed information about how to locate your book's source code, go to www.apress.com/source-code/. Readers can also access source code at SpringerLink in the Supplementary Material section for each chapter.

Printed on acid-free paper

Contents at a Glance

Contents

About the Author

Bhasker Gupta is a business analytics/data science evangelist with a proven record of incubating and jump-starting analytics at small to large organizations in addition to providing thought leadership related to the analytics/big data space to various industry players. He has more than 12 years of experience in the area of business analytics, with deep roots within the whole analytics ecosystem of India, and is widely recognized as an expert within the analytics industry.

He is the founder and editor of the hugely popular online analytics publication Analytics India Magazine (www.analyticsindiamag.com). Previously, Bhasker worked as a vice president at Goldman Sachs. He currently advises various startups on how to utilize analytics and, in his free time, thinks about interesting ideas in this field. Bhasker holds a bachelor of technology from the Indian Institute of Technology, Varanasi and a master's degree in business administration from the Indian Institute of Management Lucknow.

About the Technical Reviewer

Manoj R. Patil is a big data architect at TatvaSoft, an IT services and consulting firm. He holds a bachelor of engineering degree from the College of Engineering, Pune, India. A proven and highly skilled business intelligence professional with 17 years of information technology experience, he is a seasoned BI and big data consultant, with exposure to all the leading platforms, such as Java EE, .NET, LAMP, etc. Apart from authoring a book on Pentaho and big data, he believes in knowledge-sharing and keeps himself busy in corporate training and is a passionate teacher.
He can be reached at @manojrpatil on Twitter and at https://in.linkedin.com/in/manojrpatil on LinkedIn. Manoj would like to thank all his family, especially his two beautiful daughters, Ayushee and Ananyaa, for their patience during the review process of this book.

Acknowledgments

To my family: my parents, Dinesh Kumar Gupta and Renu Gupta; my wife, Meghna Bhardwaj; and my lovely daughter, Bani Gupta.

Introduction

Business analytics is currently the hottest and trendiest professional field, in India as well as abroad. Significant job growth is foreseen in this area in the coming years. Analytics is a multifaceted science, and a seasoned professional requires knowledge of various subjects, including statistics, database technology, IT services, and, in addition, should possess keen business acumen.

In light of the preceding, there is a dearth of academic and instructive material available in print media. We see an ever-increasing demand from budding analytics professionals for relevant questions they are likely to encounter in the course of job interviews.

This book aims to provide the necessary information to professionals preparing to interview for various job interviews related to the field of analytics.

Interview Questions in Business Analytics is a comprehensive text designed for aspiring analytics professionals or those anticipating to be interviewed for analytics/data science positions. The book targets beginners with little knowledge of analytics, as well as seasoned professionals trying to prepare for interviews or higher positions.

The focus is on in-depth concepts of data preparation, statistics, and analytics implementation, among many other crucial topics favored by interviewers.

This book follows a question-and-answer format, to help you grasp concepts quickly and clearly. Thus, it also serves as a primer for analytics. The answers to all questions are explained in detail and provide comprehensive solutions to the problems specific to the questions.

Covering more than the technical aspects of analytics, the book also addresses their business implementation. Readers will learn not only the *how* of analytics but also the *why* and *when*. Therefore, the book also targets seasoned professionals aiming to assume managerial roles within their current (or other) organizations.

CHAPTER 1

■ ■ ■

Introduction to Analytics

Data is increasingly being considered an important asset within organizations today. The lower cost of storage and vast expansion in technology have led to a deluge of data.

Organizations today think increasingly about accessing this gigantic trove of information and putting it to valuable use. Analytics is a discipline that uses different information-mining systems to glean insights from information. This is a necessary tool to aid organizations in resolving issues and achieving quick, impactful, and scientific decision making.

In this chapter, we will look at some of the basic questions regarding analytics. These will serve additionally as a brief overview of the topic. Interviewers usually ask these questions to test a prospective candidate's basic understanding of analytics.

More technical questions follow in subsequent chapters.

Q: What is analytics?

Analytics is the scientific process of transforming data into insight for better decisions.[1]

Analytics delves into data to recognize patterns that can help organizations make more informed decisions. Organizations today are sitting on a huge amount of data. Analytics helps to dig into this data to find patterns, insights, or trends that are not usually recognized otherwise. This *data mining* leads to number crunching and, finally, predications and even recommendations.

Today, technology has advanced, and we now have various tools and methods with which to analyze data that has increased the significance of analytics in organizations. Analytics has virtually changed the way businesses have been run traditionally. Gone are the days when most business insights used to come through gut feelings or the intuition of experienced managers. Today, analytics provides tangible information on which managers and executives can rely in order to make informed decisions.

---

[1]Informs, "What is Analytics," www.informs.org/About-INFORMS/What-is-Analytics, 2016.

© Bhasker Gupta 2016

B. Gupta, *Interview Questions in Business Analytics*, DOI 10.1007/978-1-4842-0599-0_1

Q: Why has analytics become so popular today?

Data crunching has been popular since a very long time. And organizations have been using data even before analytics became popular. What we are seeing today is the industrialization of data crunching and insight gathering.

We are in the midst of an explosion—a veritable data deluge, as I mentioned—and there seem to be no signs that this will change. Presently, companies have the required knowledge and tools to extract data about their customers, business units, and partners. This includes systems such as CRM, social media channels, ERP application, POS applications, and almost all IT applications, which churn out large amounts of valuable data. With the power of analytics, businesses can make use of this information, using algorithms and tools.

Data can be meaningful only when it is analyzed, shared, and made available to the relevant people. What is important is to identify what data benefits you, and at times, this decision has to be determined in nanoseconds.

Q: What has led to the explosion of analytics into the mainstream?

Two developments led to the explosion of analytics into the mainstream: cheap data storage and the availability of high-performance distributed computing. In today's business scenarios, massive amounts of data are being generated every day. With technological advancement and cheap storage capabilities, organizations find themselves collecting huge amounts of data. Every time a customer performs a transaction on an ATM, makes a retail purchase, or even visits a website, a data point is created.

Q: How is analytics used within e-commerce and marketing?

E-commerce sites heavily utilize analytics to make day-to-day decisions. Analytics evaluates visitors' data in order to make recommendations and suggest products that might interest customers. This allows for a great cross-sell opportunity. These indications are usually termed *recommendation systems* in the e-commerce industry, i.e., based on a visitor's past purchases, other items that have a high propensity of appealing to that visitor are suggested (see Figure 1-1). For example, a business may find that, historically, athletes who buy pizza and salad also purchase soft drinks. It can subsequently recommend a soft drink to any athlete who buys pizza and salad from its site.

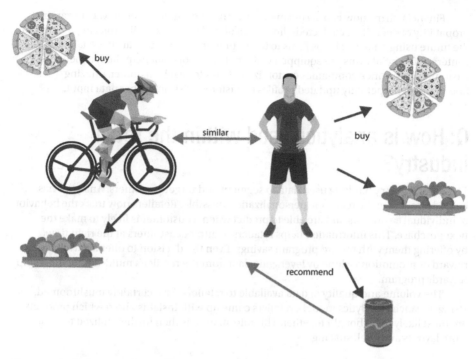

Figure 1-1. *Recommendations based on past purchases of items*

Also, using the information gleaned from specialized analytics, ads being displayed on various sites and search engines can be customized to appeal to specific visitors.

Websites rely heavily on visitors' browsing patterns to come up with site design that can more effectively entice visitors to remain on the site.

Q: How is analytics used within the financial industry?

The finance industry was the earliest adopter of analytics. Capital One gained its market-leader position in the credit card industry by being first to use analytics as a strategic differentiator. Using data from past customer preference and performance, Capital One was able to tailor-make credit card products according to the needs of each customer.

Risk analytics has gained wide acceptance in the industry, owing to its obvious benefits. By analyzing large amounts of historic customer data, financial firms can access the risk levels for each customer and, thus, that for their complete portfolios. This leads to effective fraud detection, which earlier may have remained untraceable.

Financial firms now can access many other metrics more accurately. For example, probability of default on each credit line is calculated in finer detail. Projecting losses into the future using analytics helps firms to better prepare for any unwarranted fallouts. Call centers of financial firms are equipped with analytics tools that help them to better serve customers. Insurance companies customize policies for each customer, utilizing fine-grained, perpetually updated profiles of customer risk and other data input.

Q: How is analytics used within the retail industry?

The retail industry has long used data to segment and target customers. Analytics has helped retailers make real-time personalization possible. Retailers now track the behavior of individual customers and are able to predict when a customer is likely to make the next purchase. This information helps retailers to attract customers at the right time, by offering them with reward program savings. Even the decision to offer a particular reward or promotion is done by leveraging data from the retailer's multi-tier membership rewards program.

The volume and quality of data available to retailers have certainly mushroomed. Using advanced analytics models, retailers come up with insights about which products are most likely to be bought together. This information is then further utilized to design store layouts and shelf spacing.

Q: How is analytics used in other industries?

Other sectors, too, benefit from analytics, for dividing customers into more revealing micro segments. Human resources departments access changes in work conditions and implement incentives that improve both employee satisfaction and productivity by using analytics tools. In the telecom industry, analytics is used to determine customers' calling patterns and then to offer customized products or rewards to increase customer loyalty.

Q: What are the various steps undertaken in the process of performing analytics?

The following are six broad steps in the analytics process (see Figure 1-2):

1. Business understanding
2. Data understanding and extraction
3. Data preparation and measure selection
4. Modeling
5. Model evaluation and visualization
6. Deployment

Figure 1-2. *Steps in the analytics process*

Q: Can you explain why it's important to have an understanding of business?

In addition to a knowledge of statistics and data technologies, a data scientist requires an understanding of how analytics translates to business outcomes.

Certainly, business understanding is the first and most crucial step toward applying analytics. It is important for analytics professionals to understand the state of the business that they are working in and what trends are affecting it. Also, an understanding of key business processes inside the organization is required, as well as how decisions are made within them. An analyst should focus on how analytically solving a decision-making problem will influence the business. Most of the time, analytics requires selling the advanced models inside an organization to various department stakeholders. Unless there is a clear proposition on how well analytics is solving the business problem, it will be difficult for management to buy into the solutions proposed by analytics.

Q: Why is business understanding such an integral part of analytics?

Figure 1-3 describes how organizations derive business value from analytics. Decision capabilities pretty much relate to the specific abilities analytics has to achieve for more informed decisionmaking. Thus, it is important for an analytics professional to have a domain specialist mindset. Analytic capabilities are about the diagnostic, predictive, and prescriptive abilities of analytics, and information capabilities are about describing, organizing, and integrating the data inside the organization.

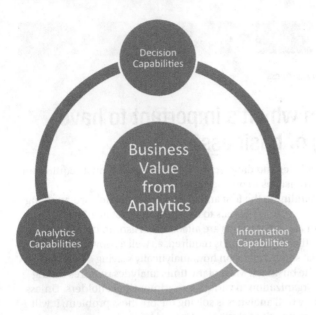

Figure 1-3. How organizations derive business value from analytics

In deriving value from business analytics, three capabilities are intertwined. While working on an analytics project, it's important to ask the following three questions: How do you leverage data for decision making? How do you leverage data for analytics? How do you leverage analytics for decision making? An analytics professional, therefore, is also a data analyst, IT professional, modeler, and domain specialist, all rolled into one.

Q: What is modeling?

An analytical model is a statistical model that is designed to perform a specific task or to predict the probability of a specific event.

In layperson's terms, a model is simply a mathematical representation of a business problem. A simple equation $y=a+bx$ can be termed as a model, with a set of predefined data input and desired output. Yet, as business problems evolve, the models grow in complexity as well. Modeling is the most complex part in the life cycle of successful analytics implementation.

Scalable and efficient modeling is critically consequential in enabling organizations to apply these techniques to increasingly voluminous data sets and for reducing the time taken to perform analyses. Thus, models must be created that implement key algorithms to determine the appropriate solution to a business quandary.

Q: What are optimization techniques?

Optimization reformats systems and processes to improve their performance. This improvement can be in terms of cost rationalization, risk mitigation, or achievement of operational efficiencies ("doing more with/in less"). Analytics is used heavily in various optimization techniques, for example, optimizing campaign performance.

Q: What is model evaluation?

Once a model is created, it is important to test its stability. Stress testing can be employed to determine the ability of a model to maintain a certain level of effectiveness under unfavorable conditions. Here, the input parameters are varied, often in simulation style, to alter the acceptable range and check the resultant effect on the model. Ideally, an effective model should continue to perform with reasonable acceptance under even extreme scenarios.

Q: What is in-sample and out-of-sample testing?

In-sample and *out-of-sample* testing particularly have gained a lot of attention recently. Suppose, for example, that we had 100 observations in a data set, and we created a model from these. We would take our model and then begin using it in production. The problem with this approach is that we would have no idea how good/reliable its predictions were, because our 100 data points would represent only one sample from the population. Another sample could have results that deviate from the model itself.

In/out-sample testing would split the sample into two parts. One part would have 80 observations, which would be our training set. We would create our model from this set and then check the reliability of the fit of this model to the 20 remaining observations. Such a test is an example of out-of-sample testing. If we were to run the model on the same 100 data sample, this would be an example of in-sample testing.

Q: What are response models?

A response model predicts the likelihood of a certain event, for example, the likelihood of a customer responding to a marketing campaign.

Q: What is model lift?

Model lift measures the efficiency of a model. It is the ratio between the results obtained after and prior to applying the model. In other words, model lift provides a measure of incremental benefit that a model brings to a situation.

For example, a company has decided to undertake an e-mail campaign to reach 100,000 of its customers. In absence of any optimization model, we can assume that the response percentage will be a historic response rate (let's say 20%), irrespective of any number of e-mails that are sent.

But, when an optimization technique is applied to the campaign, the response rate can increase with a specific number of e-mails sent out.

This can be explained using lift curves. As shown in the lift curve illustrated in Figure 1-4, the blue line is the baseline, i.e., the response rate when no model is applied. The red line is the response rate when a model is applied for optimization. The response rate increases steadily and then decreases to normal rates when 100% of customers are contacted.

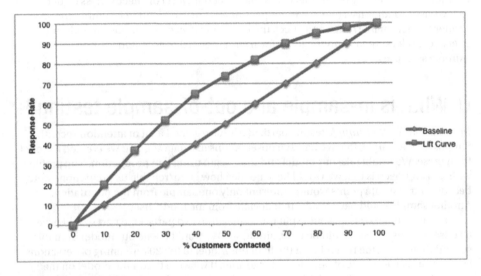

Figure 1-4. *Model lift curve*

Q: Can you explain the deployment of an analytics model? Why is it important?

Once we have a fully tested working model in place, we turn our attention to the deployment of the model within an existing IT infrastructure.

Analytics professionals usually built models on Excel or statistical tools like SAS and SPSS. The analyst can rerun these models again, yet they face the issues of scalability and availability to wider audience. For these reason, the models are deployed over more scalable business intelligence applications.

Two other issues are resolved because of model deployment. Often models require scheduled runs at certain time periods—daily, monthly, etc. Deployment of models ensures that the models are run without interventions from an analyst. Secondly, over time the model parameters change; for example, a particular regression model took into consideration one year of time series data. As we move ahead in time, the model itself might change due to availability of new or different data. A deployed model would inherently take that into consideration.

Q: How are predictive and descriptive analytics differentiated?

Descriptive analytics is used to analyze historic data and extract key insights from it. An example of the use of descriptive analytics would be to determine for a credit card company the risk profiles of a particular set of customers.

On the other hand, predictive analytics is about analyzing data to predict future outcomes. For instance, using the same example of the credit card company data, predictive analytics would predict the probability of default among the customers of the data set.

Q: How much can we rely on the results of analytics?

The basic premise of analytics is that we can guess intent, based on past behavior. Predictions from analytics should always be taken with a grain of salt. Predictive analytics assumes that the future will be similar to the past and then forecasts the past data forward. What actually occurs could certainly be very different from what has been modeled using past data.

Q: Can you briefly summarize the tools used in analytics

Excel is the most popular tool for analytics purposes. About 80% of all analytics worldwide is still being performed on Excel. SQL is widely used as an efficient data-mining tool. Most data preparation is done with SQL and later versions, and the data is exported to Excel for analysis purposes. For advanced-level analytics involving statistics, there is a wide variety of tools (both open source and commercial)available on the market. Some commonly available and often-used statistical tools include R, SAS, Python, and SPSS.

CHAPTER 2

■ ■ ■

Data Understanding

There are several things to learn about data before actually applying analytics to it. Data understanding is the first step in a robust analytics process.

We will look first at the different scales—both categorical and numeric (continuous). We will then delve deeper into data-collection methodologies and data sampling. Important aspects of data quality and sampling errors will also be discussed.

Q: What are the four types of data attributes?

A data attribute is the characteristic of an object that can vary from one object to another or from one time to another.

There are four levels (scales) of measurement that enable data scientists to systematically measure and analyze phenomena that cannot simply be counted. These levels are important to analytics professionals, because they provide guidance about the proper use of different techniques.

- *Nominal*: Defines dissimilar categories, for example, land-use classes (urban, coastal, hilly)

- *Ordinal*: Data categorized by levels or rankings, e.g., low, medium, high

- *Interval*: Data based on a scale and whose exact differences are known, e.g., temperature: 50° C is warmer than 0° C

- *Ratio*: Data similar to interval data but having a well-defined zero value. For example, a population density of zero indicates that no population exists in an area.

Nominal and ordinal data can be characterized as categorical data, whereas interval and ratio data can be grouped as numeric data (see Figure 2-1).

© Bhasker Gupta 2016
B. Gupta, *Interview Questions in Business Analytics*, DOI 10.1007/978-1-4842-0599-0_2

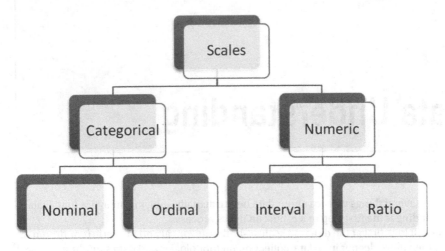

Figure 2-1. *The four types of data scales*

Q: Can you explain the nominal scale in detail?

A nominal measurement scale is used for variables in which different categories are clearly defined. These categories should be mutually exclusive and exhaustive. For example, classifying people into "male" and "female" categories.

On a nominal measurement scale, the categories do not follow any relative ordering, i.e., assigning a numerical value to the categories is arbitrary. Comparing data on a nominal scale involves only a comparison of equality, i.e., equal (=) or not equal (≠). Examples of nominal attributes include the following:

- Administrative regions
- Land-use classes (urban, coastal, hilly)
- Soil types

Q: What are some of the pitfalls associated with the nominal scale?

Questions are sometimes raised regarding whether a nominal scale should be considered a "true" scale, as assigning values to categories is only useful for categorization. Accordingly, the common notion of assigning men as "1" and women as "2" does not signify that women are "twice something or other" as compared to men. Nor do the numbers suggest any superiority—that 1 is somehow "better" than 2 (as might be the case in competitive placement).

Mathematical operations such as addition and subtraction, and statistical operations, such as mean, cannot be performed on a nominal scale. The only mathematical or statistical operation that can be performed on a nominal scale is a count or frequency.

Numbers on a nominal scale are merely for identification and nothing more.

Q: Can you explain the ordinal scale in detail?

The ordinal scale is used to order categories. This enables comparison without providing any information regarding the amount of difference. Examples of ordinal attributes are the following:

- Water quality

- Earthquake magnitudes

- Opinions on a survey

Just because numbers are assigned to categories does not mean that you can make computations from them. A lake with level 2 water quality is not necessarily twice as clean as a lake with level 4 water quality. Therefore, you cannot use a statistical mean for such data. Use the median or mode instead.

Q: What are some of the pitfalls of the ordinal scale?

Ordinal variables convey the relative difference, rather than the absolute difference, between coded values. That is, we know the relative placement of the values but have no information on how distant they are from each other.

Q: Where are ordinal scales most commonly used?

Ordinal scales are commonly used in attitudinal measurements in primary research or surveys. Five-point scales are used to rate people's perceptions about a theme under study.

Q: What is meant by data coding?

Data coding is the conversion of a nominal value to a numeric value when running statistics on them, to make it easier to compute using statistical programs.

Q: Can you explain the interval scale in detail?

An interval scale arranges values from an agreed-upon zero, or starting point, using a unit of measure. Calendar dates and temperatures measured in degrees Fahrenheit or Celsius are examples of attributes measured on interval scales. Differences can be determined (-), but sums and quotients are not possible.

The zero point and the units employed are a matter of convention and are not absolute.

Q: What are the two defining principles of an interval scale?

An interval scale measurement has two defining principles: equidistant scales and no true zero.

An equidistant scale means that the interval scale has equally distributed units.

An interval scale does not have a true zero. Temperature is a common example: even if the temperature is zero, this doesn't imply that there is no temperature. Measurements of latitude and longitude also fit this criterion. There is no value to designate the lack of the occurrence you are measuring.

Q: What are the characteristics of the ratio scale?

Simply put, a ratio scale is a scale of measurement with a defined zero point that allows us to explain the ratio comparisons. Age is an example of a ratio measurement scale. The difference between three years and five years is the same as the difference between eight years and ten years (equal intervals). Also, ten years is twice as old as five years (a ratio comparison).

A ratio scale is the product of quantification through counting, statistics, measurement, etc. The data are expressed as absolute values, sometimes including a unit. The zero point on a rational scale is not simply a convention but a true zero of the attribute. Examples of data measured on a ratio scale are population, length, area, and velocity.

A ratio scale has the characteristics of all the other three scales, making it the most refined of all.

Q: What are the limitations of scale transformation?

It is possible to transform units within scale types without losing information. From a higher to lower scale (ratio being the highest, nominal the lowest), information is lost in the transformation.

Transformations from lower scales to higher ones are not possible without additional information. For example, without more information, the names of major European rivers cannot result in a map comparing river size.

Q: In 2014, Forbes ranked Bill Gates as the richest man in the United States. According to what scale of measurement is this ranking?

Ordinal scale.

Any ranking scale is an ordinal scale, because it provides order without providing quality of difference. The difference in wealth between ranks numbered 1 and 2 might not be the same as that between ranks 2 and 3.

Q: How are continuous and discrete variables differentiated?

Variables can be continuous or discrete. A continuous variable is measured along a value range, whereas a discrete variable is measured in categories. Car speed can be timed to the nearest hundredths of a second, thus representing a continuous variable.

A discrete variable is not measured along a continuum. Most categorical variables fall into the discrete variable class.

Q: How are primary and secondary data differentiated?

Data collection is broadly divided into two types: primary data and secondary data. *Primary data* is the data observed or collected by the researcher, or on the instruction of a researcher, for a specific purpose. For example, POS (point of sale) data from a retail store is the primary data for a retailer. Also, data collected through primary research conducted by an organization via interviews, focus groups, and surveys is primary data as well.

Published or archived data collected in the past or by other parties is called *secondary data*. Examples include previously written literature reviews and case studies, census data and organizational records, published texts and statistics.

Q: Broadly, what are the four methods of primary data collection?

Following are the four methods available to a researcher whereby he or she can reach directly to a target audience and gather relevant information.

- Observation
- In-depth Interviewing
- Focus Groups
- Questionnaires and Surveys

Q: What is meant by primary data collection by observation?

The systematic notation and recording of conditions, behaviors, events, and things is called observation. For example, in so-called classroom studies, researchers carry out observations to document and describe complex actions and interactions.

Q: What is primary data collection by in-depth interviewing?

In-depth interviews, on which qualitative researchers rely quite extensively, are conversations conducted with a set audience given a predefined questionnaire, in order to gauge their perception of a topic.

Q: What is primary data collection by focus groups?

Interviewing focus groups has been widely adopted for primary data gathering. The groups generally consist of participants who are asked focused questions to encourage discussion. These interviews may be conducted several times with different individuals, to help the researcher identify accurate insights.

Q: What is primary data collection by questionnaires and surveys?

Questionnaires consist of structured questions with response categories. Sometimes open-ended questions are also included in a questionnaire. The critical assumption is that self-reporting can help to measure or describe characteristics or beliefs.

Sample surveys are most apt when a researcher has to draw inferences about a large group based on a small sample of a population. It helps to describe and explain the properties, features, and variability in a population statistically.

Q: What is data sampling?

Sampling is the process of selecting a small set of a whole population. It is often undertaken, as it's cheap (it's less expensive to sample 1,000 sports fans than 100 million) and practical (e.g., performing a screen break test on every mobile phone is unfeasible).

Based on a very small sample compared to an entire population, researchers can draw inferences about a population factor via statistical inference. A larger sample size is expected to yield more accurate sample estimates.

Q: What are the different types of sampling plans?

A sampling plan is the method according to which a sample will be drawn from a population. Following are the three common sampling methods:

- Simple random sampling
- Stratified random sampling
- Cluster sampling

Q: What is simple random sampling?

Random sampling is by far the most common method used. A simple random sample is the method by which every possible sample has the probability of getting selected.

Drawing three balls from a box containing balls of different colors is an example. The probability of selecting any group of three colored balls is the same as picking three colored balls from any other group.

Q: What is stratified random sampling?

A stratified random sample is a two-staged process achieved first by creating mutually exclusive strata from a population and then taking simple random samples from each stratum. For example, to sample a retailer's customers, you should first segment the customers into demographic groups, such as age groups, and then randomly sample each segmented group equally.

One advantage of stratification is that not only can we obtain information about an entire population, but inferences about each strata can also be drawn.

Q: What is cluster sampling?

A cluster sample is also a two-staged sampling method whereby a simple random sample is used to create clusters (groups) of elements. Simple random sampling is subsequently performed, to select elements from clusters.

Q: What are the different errors involved in sampling?

The accuracy of sample estimates increases by increasing the sample size. Two major types of errors associated with sampling are

- Sampling errors
- Non-sampling errors

Q: What causes a sampling error?

A sampling error is caused as a result of the observations selected for the sample. The error causes differences between the sample estimates and the population estimates. Sampling errors are random, and we have no control over them.

For example, the true average height of all men in the universe and the average height computed by a sample of men would show some difference. This difference is due to sampling error and cannot be controlled, given a sample.

Q: What is a non-sampling error?

Non-sampling errors are caused by errors made during data acquisition or due to improper sample observations.

Q: What are the three types of non-sampling errors?

The three types of non-sampling errors are

- Errors in data acquisition
- Nonresponse errors
- Selection bias

Q: Can you identify some errors in data acquisition?

Errors in data collection may arise from the recording of incorrect answers, owing to the following:

- Faulty equipment, causing errors in measurement
- Errors in the primary sources
- Misinterpretation of terms, resulting in inaccurate data being recorded
- Participants giving inaccurate responses to questions concerning sensitive issues

Q: When does selection bias occur?

Selection bias occurs when some member of a population cannot be selected because of the sampling plan, for example, while calculating the average height of males in an area, the researcher does not select the top-ten tallest males in his sample.

Q: What is data quality?

Data quality checks or assesses if the data is fit to serve the purpose of being collected for a given context.

Q: What is data quality assurance?

The procedure of inspecting the consistency and efficacy of data is called data quality assurance (DQA).

Periodical cleaning and sorting through the data is required for maintaining data quality. This typically involves bringing it up to date, normalizing it, and de-duplicating it.

Q: What are various aspects of data quality?

Aspects of data quality include the following (see Figure 2-2):

- Completeness
- Accuracy
- Consistency
- Timeliness
- Auditability
- Non-redundancy
- Integrated

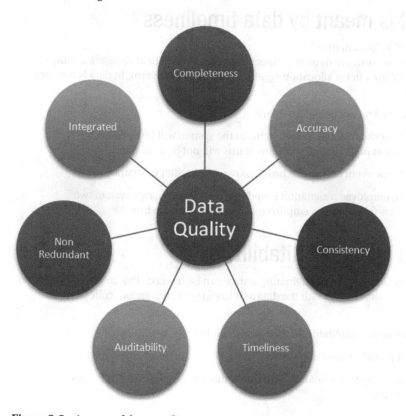

Figure 2-2. *Aspects of data quality*

Q: What is data consistency?

Data across an enterprise should be consistent.
Examples of data inconsistency include

- An active e-mail account of an employee who has resigned

- A canceled and inactive mortgage, although the mortgage billing status indicates "due"

Accurate data (i.e., data that correctly represents actuality) can still be inconsistent.

- An online promotion campaign with a January 31 closing date might show a booking under February 2.

Data can be complete but inconsistent.

- Data for all completed flights from New York to Chicago may be available, but the status of some flights indicates "to take off."

Q: What is meant by data timeliness?

"Data delayed" is "data denied."
Data timeliness depends on the expectations of the user. Real-time availability of data is required for a ticket allocation system in aviation, but overnight data is fine for a billing system.

Examples of data not being timely are

- A check is cleared, but its status in the system will be updated only at night, hence, real-time status will not be available.

- A new event becomes a headline two weeks after it occurs.

- An employee resignation is updated on the company's system two years following the employee's departure from the firm.

Q: What is data auditability?

Auditability means that data's originating source can be tracked. This requires a common identifier, which should stay with the data as it undergoes conversion, collection, and reportage.

Examples of non-auditable data include the following:

- A provident fund account that cannot be linked to an employee ID

- An entry into a financial statement that cannot be linked to a sales invoice

Auditable data includes such examples as a provident fund account being associated with an employee ID or a transaction in a financial statement being accompanied by a relevant sales invoice.

Q: What is meant by data being "redundant"?

Data redundancy refers to data duplication in more than one place or more than one form. There might be valid reasons for data being redundant, however, for example:

- The redundant data was created because the duplicated data provides enhanced availability or performance characteristics.

- Part of the data should not be exposed to some viewers.

- Production and development data are created separately.

In these cases, redundant data is only the symptom and not the cause of the problem. Only managerial vision, direction, and robust data architecture leads to an environment with less redundant data.

Q: What is data cleaning?

Data cleaning is the process of error and discrepancies removal from data, to increase the quality of data.

Removing outliers or filling up holes in data are examples of data-cleaning processes.

Q: What are different sources of errors in data?

Errors in data can creep in during steps involving both human interaction and computation. The sources of data error fall into the following categories:

- Data entry errors

- Measurement errors

- Distillation errors

- Data integration errors

Q: What are data entry errors?

Data entry is mostly done by humans, who typically extract information from speech (e.g., in call centers) or enter data from written or printed sources. In these instances, the data may have typographic errors inserted during entry, or data is entered incorrectly, due to misunderstanding of the data source, e.g., a call center employee records wrong information about a customer on a call, due to poor reception.

Q: What are distillation errors?

Data distillation is the preprocessing and summarization of data before being entered into a database. Distillation is undertaken either to reduce the size of data or to remove redundant data.

Data distillation has the likelihood of producing errors in the data. The distillation process itself can have errors in it, so that post-distillation of the data might not be truly representative of the information stored earlier.

Q: What are data integration errors?

Big databases are generally collected from different sources over time. The data integration procedure from multiple sources can lead to errors. The final data post-integration might not be a true representation of the original data or may not contain the same information as was entered earlier.

Q: What is data parsing?

Parsing in data cleansing is performed to detect syntax errors. One parses through a database to identify entries that do not adhere to the syntax defined in the parser.

Q: What are data outliers?

Outliers are observations that are extreme, relative to other observations observed under the same conditions.

It's an unfortunate fact of research that data are not always well-mannered. "Outliers" result in almost all research projects involving data collection. Data-entry errors or rare events (such as readings from a thermometer left near a furnace, a change in accounting practice, or large fingers)—all these (and many more) explain the presence of outliers in a collection of data.

Q: What are the sources of outliers?

The three main sources of outliers in data are

- Data-entry errors
- Implausible values
- "Rare" events

CHAPTER 3

■ ■ ■

Introduction to Basic Statistics

Statistics is an amalgamation of both science and art. It is built around the rules of applied mathematics, which is the science part. For quantitative research, it requires rigorous statistical analysis. Statistical calculations are performed under a set of rules, to minimize incorrect results. The art part of statistics is the judgment of the practitioner. It requires judgment and decision making in deciding how to design a problem statement and how to solve it by using data and statistics. Practitioners must decide among the various statistical techniques to use at what significance level and then interpret the produced results.

In this chapter, I have provided questions that are commonly asked in regard to the concept of probability and its associated rules of addition and multiplication, to solve a range of likely problems. The notion of a discrete and continuous probability distribution is introduced, and examples are provided, to illustrate the different types of distributions.

Q: How would you define statistics to a layperson?

Statistics is a set of tools used to organize and analyze data. When referring to statistics, one usually means one or more of the following:

- A set of numerical data, such as the unemployment rate of a city, annual number of deaths due to bee stings, or the breakdown of the population of a city according to race in 2006 as compared to 1906

- Numbers or statistical values such as mean, which describes a sample of data, rather than the whole population

- Results of statistical procedures such as z-test or chi-square statistics

- A field of study using mathematical procedures to make inferences from data and decisions based on it

© Bhasker Gupta 2016
B. Gupta, *Interview Questions in Business Analytics*, DOI 10.1007/978-1-4842-0599-0_3

Q: How are statistical methods categorized?

There are two categories:

- Descriptive statistics
- Inferential statistics

Q: What is the definition of descriptive statistics?

Descriptive statistics is the part of statistics that summarizes important characteristics of data sets. It is used to derive useful information from a set of numerical data.

Q: How do you define inferential statistics?

Inferential statistics consists of drawing conclusions, forecasts, judgments, or estimates about a larger data set, using the statistical characteristics of a sample of data.

Q: What is a population data set?

A population data set refers to all elements of the set (or the whole universe of the set items), such as the income of each person in a country or the daily sales of a product across every store of an area. A census is the data set having the desired information for all items in the population. In real life, we usually have data from a sample of the population, such as the last daily sales data for the previous two years.

Statistics is the number used to define a sample data set from which we can estimate a population parameter. For example, the mean of a sample data set (X or $\mu^\wedge$) provides the estimate of population mean μ.

Q: What are data frequency tables?

Data frequency tables are a method of representing a summary of data in a way in which identifying patterns or relationships becomes easier. By properly representing a data set in data frequency tables, we can unlock a lot of information about the data. It shows how many times something occurs in a given data set and how the data is distributed.

Q: How do you define location statistics?

Location statistics (the measure of a central tendency) is the central point of the data set. This statistic can be used to signify the representative (expected) value of a data set.

Q: What is the mean or average?

Mean (or average) is the numerical value of the center of a distribution.

$$\text{Mean} = \frac{X_1 + X_2 + X_3 + \ldots\ldots + X_n}{n}$$

■ **Note**   Mean is also referred to as $\bar{X}$.

The mean of a sample of data is typically denoted by $\bar{X}$, whereas the Greek letter μ (mu) is used to symbolize the mean of an entire population.

Q: What are some of the key characteristics of a mean?

The following are key characteristics of a mean:

- Each data set, be it interval or ratio, has an arithmetic mean.

- Arithmetic mean computation includes all the data values.

- The arithmetic mean is unique.

Outliers (unusually large or small values) affect the calculated value of an arithmetic mean. The mean of 2, 5, 10, and 120 is 34.25, which does not provide a clear picture about the data. However, while calculating the arithmetic mean, all the observations are considered, which is a positive thing.

Q: What is a weighted mean?

When each observation disproportionately influences the mean, *weighted mean* is calculated. This can be caused either by outliers, or the data is structured in such a way that there are clusters within data groups that have varied characteristics.

For a set of numbers, the following equation is used to calculate weighted mean:

$$\bar{X}w = \sum_{i=1}^{n} w_i X_i = \left(w_1 X_1 + w_2 X_2 + \ldots\ldots + w_n X_n \right)$$

where:

$$X_1, X_2, \ldots\ldots X_n = observed values$$

$$w_1, w_2, \ldots\ldots w_n = corresponding weights$$

25

Q: What is a median? How do we determine the median of a data set?

Median (also known as the 50th percentile) is the middle observation in a data set. Median is calculated by sorting the data, followed by the selection of the middle value. The median of a data set having an odd number of observations is the observation number [N + 1] / 2. For data sets having an even number of observations, the median is midway between the observation numbers N / 2 and [N / 2] + 1. N is the number of observations.

As shown in Figure 3-1, the median is the middle observation point in a data set.

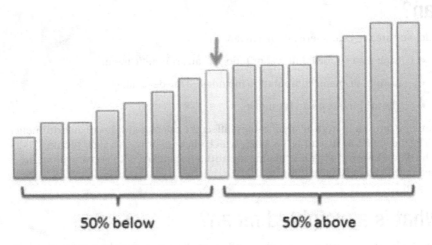

Figure 3-1. The median in a data set

Q: When is a median used as opposed to a mean?

In the case of outliers (extreme values) or of a skewed data set (when one tail is significantly longer in a bell-shaped curve), the median is more applicable. If you want to represent the center of a distribution, such as in the case of the salaries of ten employees and one CEO, when the CEO has a significantly higher salary, using a median is more appropriate.

Q: What is meant by "mode"?

The value appearing most frequently in a data set is called the mode. A data set may have single or multiple modes, referred to as unimodal, bimodal, or trimodal, depending on the number of modes.

Q: Can you define quartiles, quintiles, deciles, and percentiles?

Quantile is the broad term for a value that divides a distribution into groups of equal size. For example:

- *Quartile*: Distribution separated into four equal intervals. This is detailed in Figure 3-2.

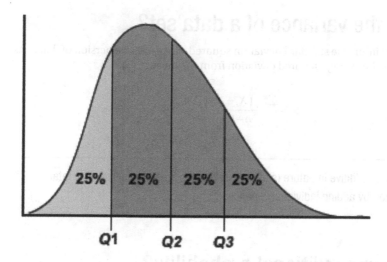

Figure 3-2. *Quartiles*

- *Quintile*: Distribution separated into five equal intervals
- *Decile*: Distribution separated into ten equal intervals
- *Percentile*: Distribution separated into 100 equal intervals (percent)

Q: Can you define standard deviation? Also, can you discuss briefly the notations used for it?

Standard deviation (StDev) is the measure of deviation of all observations in a distribution from the mean. It is also known as root-mean-square, for a sample data set calculated using the following formula.

$$S = \sqrt{\frac{\sum_{i=0}^{n} \left(X_i - \bar{X}\right)^2}{n-1}}$$

S is used to signify the sample standard deviation, whereas the Greek letter σ signifies the population standard deviation. The terms S or $\sigma^{\wedge}$ signify the estimated population standard deviation.

Q: What is the variance of a data set?

Variance, which is simply the standard deviation squared, represents dispersion of data. In other words, it is the average squared deviation from the mean.

$$S^2 = \frac{\sum_{i=0}^{n} \left(X_i - \bar{X}\right)^2}{n-1}$$

■ **Note**   Variance is additive in nature (standard deviations are not additive). The total variance is computed by adding individual variances.

Q: What is unconditional probability?

Unconditional probability is the probability of an event irrespective of occurrences of other events in the past or future. For example, to find the probability of an economic recession, we calculate the unconditional probability of recession only, regardless of the changes in other factors, such as industrial output or consumer confidence.

Unconditional probability is also known as marginal probability.

Q: What is conditional probability?

Conditional probability results when two events are related, such that one event occurs only when another event also occurs. Conditional probability is expressed as P(A|B); the vertical bar (|) indicates "given," or "conditional upon."

The probability of a recession when consumer confidence goes down is an example of conditional probability expressed as P (recession | decrease in consumer confidence). A conditional probability is also called its likelihood.

Q: What is the multiplication rule of probability?

Joint probability is the probability of two events happening together. The calculation of joint probability is based on the multiplication rule of probability and is expressed as

$$P(AB) = P(A|B) \times P(B)$$

This is read as follows: the conditional probability of A given B, P(A|B), multiplied by the unconditional probability of B, P(B), is the joint probability of A and B.

Q: What is Bayes's theorem?

Bayes's theorem is one of the most common applications of conditional probability. A typical use of Bayes's theorem in the medical field is to calculate the probability of a person who tests positive on a screening test for a particular disease actually having the disease. Bayes's formula also uses several of the basic concepts of probability introduced previously and is, therefore, a good review for the entire chapter. Bayes's formula for any two events, A and B, is

$$P(A|B) = \frac{P(A \& B)}{P(B)} = \frac{P(B|A)P(A)}{P(B|A)P(A) + P(B|{\sim}A)P({\sim}A)}$$

You would use this formula when you know P(B|A) but want to know P(A|B).

Q: How would you define a symmetrical distribution?

A symmetrical distribution has an identical shape on each side of its mean (see Figure 3-3). This symmetry implies that on either side of the mean, the intervals will indicate the same frequency.

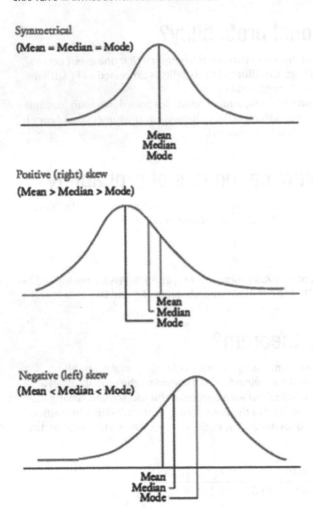

Figure 3-3. *Skewness of a distribution*

Q: What is meant by the "skewness" of a distribution? Also, what is positive and negative skewness?

Skewness refers to the degree to which a distribution is not symmetrical. Skewness can be positive or negative, depending on the existence of outliers.

Q: What are outliers? How do they affect the skewness of a distribution?

Outliers are data points having extremely large positive or negative values, compared to the rest of the data points.

When the outliers lie in the upper region, or right tail, elongating this tail, they form a positively skewed distribution. When the outliers lie in the lower region, or left tail, elongating this tail, they form a negatively skewed distribution.

Q: How does skewness affect the location of the mean, median, and mode of a distribution?

For symmetrical distribution,

Mode = Median = Mean

For a positively skewed distribution,

Mode < Median < Mean

This is illustrated in the preceding Figure 3-3. In the case of a positively skewed distribution, the positive outliers pull the mean upward, or affect the mean positively. For example, a student scoring 0 on an exam, when all other students have scored above 50, pulls down the class average, making it skew negatively.

For a negatively skewed distribution,

Mean < Median < Mode

The negative outliers pull the median down, or to the left.

Q: What is a normal distribution?

This is a bell-shaped frequency distribution having the mean value in the middle of a demonstrable number of properties (see Figure 3-4).

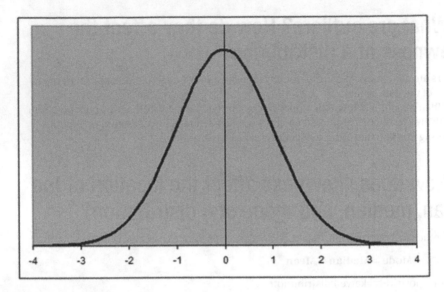

Figure 3-4. *Example of a normal distribution curve*

The total area of a normal curve is always 1, with the area on either side of the mean equal to 0.5. The area under the curve represents the probabilities.

An important attribute of the normal distribution curve is the way distribution is spread around the mean. We can see from Figure 3-5 that 68% of the population's values lie within $\mu \pm \sigma$, 95% of the population's values lie within $\mu \pm 1.96\sigma$, and 99% of population's values lie within $\mu \pm 2.58\sigma$.

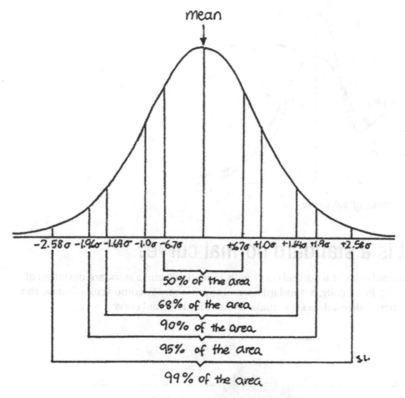

Figure 3-5. *% of area covered by graph around mean*

Q: What is the kurtosis of a distribution? Also, what are the various types of kurtosis?

Kurtosis is a measure of how high or low a distribution is "peaked" compared to a normal distribution. A distribution that is peaked more than the normal curve is described as being *leptokurtic,* whereas a distribution flatter than the normal curve is called *platykurtic.* A distribution is *mesokurtic* if it has the same kurtosis as a normal distribution. (See Figure 3-6.)

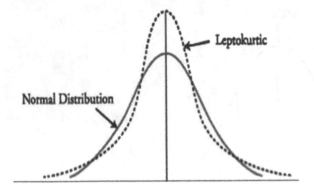

Figure 3-6. *Kurtosis of a distribution*

Q: What is a standard normal curve?

A *standard normal curve* is a normal curve with a mean of 0 and a standard deviation of 1 (see Figure 3-7). Practically, a standard normal curve is rare to come across, but we can transform or standardize a data set to make it a standard normal curve.

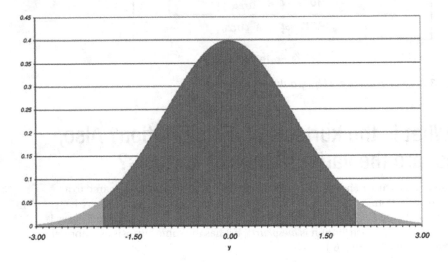

Figure 3-7. *A standard normal curve*

Q: What are some of the other continuous probability distributions?

Some commonly used continuous probability distributions are

- Student's t-distribution

- Chi-square distribution

- F distribution

Q: What is an F distribution?

An F distribution is used to test whether the ratios of two variances from normally distributed statistics are statistically different. In principle, this is the probability distribution associated with F statistics.

Q: What is a binomial probability distribution?

Experiments having only two outcomes—"success" or "failure"—such as flipping a coin to determine heads or tails, an out or not out in baseball or cricket, a person being dead or alive, a goal or no goal in soccer. These outcomes are labeled as "success" or "failure," or 1 or 0. Note that this carries no implication of "goodness."

CHAPTER 4

■ ■ ■

Hypothesis Testing

Hypothesis testing forms the next building block in learning statistical techniques. Now that you are familiar with probability distributions, the next step is to validate a data point or whether a sample falls into these distributions. Building a hypothesis is the first step in conducting an experiment or designing a survey. Don't just look on hypothesis testing as a statistical technique, but try to understand the core principles of this concept.

Q: What is a hypothesis?

It is a supposition or assertion made about the world. A hypothesis is the starting point of an experiment, in which an assertion is made about some available data, and further investigation will be conducted to test if that assertion is correct or not.

Q: What is hypothesis testing?

It is the process in which statistical tests are used to check whether or not a hypothesis is true, using data. Based on hypothetical testing, we choose to accept or reject a hypothesis.

An example: Research is being conducted on the effect of TV viewing on obesity in children. A hypothesis for this would be that children viewing more than a certain amount of hours of television are obese. Data is then collected, and hypothesis testing is done to determine whether the hypothesis is correct or not.

Q: Why is hypothesis testing necessary?

When an event occurs, it can be the result of a trend, or it can occur by chance. To check whether the event is the result of a significant occurrence or merely of chance, hypothesis testing must be applied. In the preceding example of TV viewing and obesity, the hypothesis may be incorrect, and the data may show that it is merely chance that watching television makes some children obese.

© Bhasker Gupta 2016
B. Gupta, *Interview Questions in Business Analytics*, DOI 10.1007/978-1-4842-0599-0_4

Q: What are the criteria to consider when developing a good hypothesis?

A hypothesis is the initial part of a research study. If the hypothesis formed is incorrect, the research study is also likely to be incorrect; therefore, it should be properly considered and contemplated and should include the following criteria:

- The hypothesis should be logically consistent and make sense with regard to literature and language.

- The hypothesis should be testable. If a hypothesis cannot be tested, it has no use.

- It should be simple and clear, to avoid possible confusion.

Q: How is hypothesis testing performed?

There are several statistical tests available for hypothesis testing. The first step is to formulate a probability model based on the hypothesis. The probability model is also decided on the basis of the data available and the informed judgment of the researcher. Then, depending on the answers required, the appropriate statistical tests are selected.

Q: What are the various steps of hypothesis testing?

Hypothesis testing is conducted in four steps.

1. Identification of the hypothesis needed to be tested, for example, research to check the obesity in teenagers.

2. Selection of the criterion upon which a decision as to whether the hypothesis is correct or not is to be taken. For example, in the preceding problem, the criterion could be the body mass index (BMI) of the teenagers.

3. Determining from the random sample the statistics we are interested in. We select a random sample and calculate the mean. For example, a random sample of 1,000 teenagers is selected from a population, and their mean BMI is calculated.

4. Compare the result with the expected result, to check the validity. The discrepancy between the expected and real result helps to decide whether the claim is true or false.

Q: What is the role of sample size in analytics?

Sample size for a statistical test is very important. Sample size is inversely proportional to standard error, i.e., the larger the sample size, the lesser the standard error and the greater the reliability. However, larger sample size means that a very small difference can become statistically significant, which may not be clinically or medically significant. The two main aspects of any study are generalizability (external validity) and validity (internal validity). Large samples have generalizability but not validity aspects.

Q: What is standard error?

The standard error (denoted by σ) is the standard deviation of a statistic. It reflects the variation caused by sampling. It is inversely proportional to sample size.

Q: What are null and alternate hypotheses?

A null hypothesis is the statement about a statistic in a population that is assumed to be true. It is the starting point of any research study. Based on statistical tests, a decision is taken as to whether the assumption is right or wrong.

An alternative hypothesis is the contradictory statement that states what is wrong with the null hypothesis.

We test the validity of a null hypothesis and not of an alternative hypothesis. An alternative hypothesis is accepted when the null hypothesis is proved to be wrong.

An example would be a study conducted to determine the mean height of a class of students. The researcher believes that the mean height is 170 centimeters (cms). In this case,

H_0: μ_{height} = 170 cms

H_A: μ_{height} ≠ 170 cms

Q: Why are null and alternate hypotheses necessary?

Following are the reasons null and alternate hypotheses are necessary:

- The two hypotheses provide a rough explanation of the occurrences.

- They provide a statement to the researcher that acts as the base in a research study and is directly tested.

- They provide the outline for reporting the interpretations of the study.

- They behave as a working instrument of the theory.

- They verify whether or not the test is maintained and is detached from the investigator's individual standards and choices.

Q: How are the results of null/alternate hypotheses interpreted?

Statistical tests are conducted to check the validity of null hypotheses. When a null hypothesis is proved to be wrong, the alternate hypothesis is accepted. Consider, for example, a courtroom scenario. When a defendant is brought to trial, a null hypothesis is that he is innocent. The jury considers the evidence to decide whether or not the defendant is guilty.

In the preceding courtroom example, if there is insufficient evidence, the jury will free the defendant rather than convicting him or her. Similarly, in statistics, a null hypothesis is accepted if the research fails to prove otherwise, rather than endorsing an alternative hypothesis.

Q: What is meant by "level of significance"?

Level of significance is the criteria by which a decision is reached. In the courtroom example, the level of significance can be stated as the minimum level of evidence required by the jury to reach a verdict regarding the guilt or innocence of the defendant. Similarly, in statistics, it is the criterion by which a null hypothesis is rejected. Level of significance is denoted by α.

Where to establish the level of significance is determined by the alternative hypothesis. If the null hypothesis is true, the sample mean is equal to the mean population on average. If α = 5%, it means 95% of all the sample means lie within the range of μ±σ. Let us consider an example in which the null hypothesis is that in the United States children watch three hours of TV. The level of significance is set at 95%. The other 5% value lies outside the range of μ±σ. The alternative hypothesis states that the children do not watch three hours of TV (either more or less).

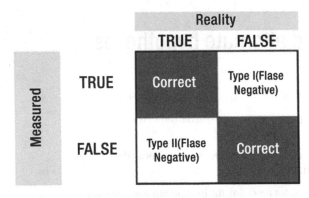

Figure 4-1. *Type I & Type II errors*

If a sample has a mean of four hours, we will calculate the outcome by determining its likelihood. We can see, then, how far the number of standard deviations for this result is from the mean. If the significance level is decided at 95%, and the distance from mean is more than one standard deviation from the mean, it implies that the null hypothesis is true.

Q: What is test statistics?

Test statistics refers to a mathematical formula determining the likelihood of finding sample outcomes, if the null hypothesis is true, to make a decision regarding the null hypothesis.

If the level of significance is set at 95% and the test statistic value is less than 0.05, this would mean that the null hypothesis is wrong and should be rejected. Therefore, the researcher can take either of the following two decisions:

- Reject the null hypothesis, when the test statistic is less than α.

- Retain the null hypothesis, when the test statistic is greater than α.

Q: What are the different types of errors in hypothesis testing?

When we perform hypothesis testing, there can be errors such as falsely accepting or rejecting the null hypothesis. There are two types of distinguished errors: *type I errors* and *type II errors*. Refer to Figure 4-1 for more details.

A *type I error* occurs when a null hypothesis is incorrectly rejected and an alternate hypothesis is accepted. The type I error rate or significance level is denoted by α. It is generally set at 5%. In the courtroom example, if the judge convicts an innocent defendant, he/she is committing a type I error.

A *type II error*, or error of second kind, occurs when a null hypothesis is incorrectly accepted when the alternate hypothesis is true. If a type I error is a case of a false positive, a type II error is a case or a false negative. It is denoted by β and is related to the power of a test. In the example of the courtroom trial, if a judge lets a guilty defendant free, he is committing a type II error.

Q: What is meant by the statement "A result was said to be statistically significant at the 5% level."?

The result would be unexpected if the null hypothesis were true. In other words, we reject the null hypothesis.

Q: What are parametric and non-parametric tests?

In a parametric statistical test, assumptions such as that a population is normally distributed or has an equal-interval scale are made about the parameters (defining properties) of the population distribution. A non-parametric test is one that makes no such assumptions.

Q: What differentiates a paired vs. an unpaired test?

When we are comparing two groups, we have to decide whether to perform a paired test. A repeated-measures test, as it is called, is used when comparing three or more groups.

When the individual values are unpaired or matched or related among one another between groups, we use an unpaired test. In cases in which before and after effects of a study are required, a paired or repeated-measures test is used. In the case of measurements on matched/paired subjects, or in one of repeated lab experiments at dissimilar times, each with its own control, paired or repeated-measures tests are also used.

Paired tests are selected for closely correlated groups. The pairing can't be based on the data being analyzed, but before the data were collected, when the subjects were matched or paired.

Q: What is a chi-square test?

A chi-square (χ^2) test is used to examine if two distributions of categorical variables are significantly different from each other. Categorical variables are the variables in which the value is in a category and not continuous, such as yes and no or high, low, and medium or red, green, yellow, and blue. Variables such as age and grade-point average (GPA) are numerical, meaning they can be continuous or discrete. The hypothesis for a χ^2 test follows:

H_0: There is no association between the variables.

H_A: There is association between them.

The type of association is not specified by the alternative hypothesis though. So interpretation of the test requires closer attention to the data.

Q: What is a t-test?

A t-test is a popular statistical test to draw inferences about single means or about two means or variances, to check if the two groups' means are statistically different from each other, where n < 30 and the standard deviation is unknown.

Figure 4-2 here shows idealized distributions. As represented in the figure, the means of the control and treatment group will most likely be located at different positions. The t-test checks if the means are statistically different for the two groups.

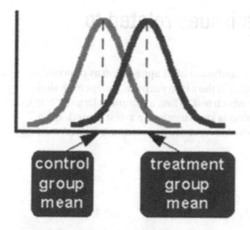

Figure 4-2. *Idealized distributions*

The t-test judges the difference between the means relative to the spread or variability of the scores of the two groups.

Q: What is a one-sample t-test?

A one-sample t-test compares the mean of a sample to a given value, usually the population mean or a standard value. Basically, it compares the observed average (sample average) with the expected average (population average or standard value), adjusting the value of the number of cases and the standard deviation.

Q: What is a two-sample t-test?

The purpose of the two-sample t-test is to determine if two population means are significantly different. The test is also known as the independent samples t-test, since the two samples are not related to each other and can therefore be used to implement a between-subjects design. In addition to the assumption of independence, both distributions must be normal, and the population variances must be equal (i.e., homogeneous).

Q: What is a paired-sample t-test?

The purpose of the repeated-measures t-test (or paired-sample t-test) is to test the same experimental units under different treatment conditions—usually experimental and control—to determine the treatment effect, allowing units to act as their own controls. This is also known as the dependent samples t-test, because the two samples are related to each other, thus implementing a within-subjects design. The other requirement is that sample sizes be equal, which is not the case for a two-sample t-test.

43

Q: Briefly, what are some issues related to t-tests?

The biggest issue with t-tests results from the confusion of its application as opposed to the z-test. Both statistical tests are used for almost the same purpose, except for a slight difference; the difference being when to use which test. When a sample is large ($n \geq 30$), and whether the population standard deviation is known or not, a z-test is used. For a limited sample ($n < 30$), when the standard deviation of the population is unknown, a t-test is chosen.

CHAPTER 5

■ ■ ■

Correlation and Regression

This chapter is concerned with measuring the relatedness between two variables. A simple measure, the correlation coefficient, is commonly used to quantify the degree of relationship between two variables. In this chapter, I will discuss different types of regression models, assumptions and questions related to them, and the estimation method used commonly with regression models.

Q: What is correlation and what does it do?

In analytics, we try to find relationships and associations among various events. In the probabilistic context, we determine the relationships between variables. Correlation is a method by which to calculate the relationship between two variables. The coefficient of a correlation is a numerical measure of the relationship between paired observations (X_i, Y_i), $i = 1, ..., n$. For different coefficients of correlations, the relationship between variables and their interpretation varies.

There are a number of techniques that have been developed to quantify the association between variables of different scales (nominal, ordinal, interval, and ratio), including the following:

- *Pearson product-moment correlation* (both variables are measured on an interval or ratio scale)

- *Spearman rank-order correlation* (both variables are measured on an ordinal scale)

- *Phi correlation* (both variables are measured on a nominal/dichotomous scale)

- *Point biserial* (one variable is measured on a nominal/dichotomous scale, and one is measured on an interval or ratio scale)

© Bhasker Gupta 2016
B. Gupta, *Interview Questions in Business Analytics*, DOI 10.1007/978-1-4842-0599-0_5

Q: When should correlation be used or not used?

Correlation is a good indicator of how two variables are related. It is a good metric to look at during the early phases of research or analysis. Beyond a certain point, however, correlation is of little use.

A dip in a country's gross domestic product (GDP), for example, would lead to an increase in the unemployment rate. A casual look at the correlation between these two variables would indicate that there is a strong relationship between them.

Yet, the extent or measure of this relationship cannot be ascertained through a correlation. A correlation of 75% between two variables would not mean that the two variables are related to each other by a measure of .75 times.

This brings us to another issue that is often misunderstood by analysts. Correlation does not mean causation. A strong correlation between two variables does not necessarily imply that one variable causes another variable to occur.

In our GDP vs. unemployment rate example, this might be true, i.e., a lower GDP rate might increase unemployment. But we cannot and should not infer this from a correlation. It should be left to the sound judgment of a competent researcher.

Q: What is the Pearson product-moment correlation coefficient?

It is easy to determine whether two variables are correlated, simply by looking at the scatter plot (each variable on the 2 axis of the plot). Essentially, the values should be scattered across a straight line of the plot, in order for the two variables to have a strong correlation.

However, to quantify the correlation, we use the Pearson's product-moment correlation coefficient for samples, otherwise known as Pearson's r.

The correlation can be either positive or negative.

So, the value of ρ can be any number between -1 to 1, as follows:

➔ $[-1 <= \rho <= +1]$

A correlation coefficient of less than zero would mean that the increase of one variable generally leads to the decrease of the other variable. A coefficient greater than zero implies that an increase of one variable leads to an increase in the other variable.

Higher values mean stronger relationships (positive or negative), and values closer to zero depict weak relationships. A correlation of 1.00 means that the two values are completely or perfectly positively correlated; -1.00 means that they are perfectly negatively correlated; and a correlation of 0.00 means that there is no relationship between the two variables.

Q: What is the formula for calculating the correlation coefficient?

The formula is as follows:

$$r = \frac{\sum\limits_{i=1}^{n}\left(\dfrac{x_i - \bar{x}}{s_x}\right)\left(\dfrac{y_i - \bar{y}}{s_y}\right)}{n-1}$$

To compute r, the following algorithm, corresponding to the preceding formula, is used as follows:

- For each (x, y) set of coordinates, subtract the mean from each observation for x and y.

- Divide by the corresponding standard deviation.

- Multiply the two results together.

- The result is then added to a sum.

- The sum is divided by the degrees of freedom, n - 1.

Q: Briefly, what are the other techniques for calculating correlation?

Although the Pearson product-moment correlation is the most widely used correlation technique, other correlation techniques must be applied if the main tenet of the Pearson technique is violated, i.e., both variables should be on an interval or ratio scale.

Spearman Rank-Order Correlation

This techniqe is used when both variables are on an ordinal scale. Both variables are ranked manually between a value of 1 to 10. The main tenet here is that the rank for a particular observation should be high for both variables, for any correlation to exist.

Phi Correlation

This is also used as an after test for a chi-square test. It is used when variables are nominal.

Point Biserial

This method is used when one variable is on a nominal/dichotomous scale and one is measured on an interval or ratio scale.

Q: How would you use a graph to illustrate and interpret a correlation coefficient?

Visualization helps us to understand the relationships among variables better. In Figure 5-1, graphs have been plotted to illustrate and visualize relationships between the variables x and y that are being statistically analyzed.

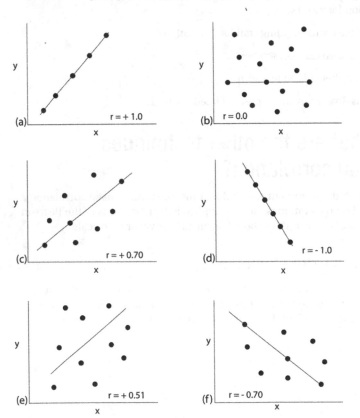

Figure 5-1. Graphs plotting the statistical analysis of relationships between the variables x and y

Now let us interpret these graphs.

```
Figure (a)
r = 1.0
```

This represents a perfect linear association. All the data points fall on the line.

```
Figure (b)
r = 0
```

No linear relationship exists between the variables. The data points are scattered randomly and may approximate a circle. Changing the value of one variable has no effect on the value of the other.

```
Figure (c)
r = 0.70
```

There is some positive linear relationship, although it is not perfect. Most of the data points fall on or closer to a straight line.

```
Figure (d)
r = -1.0
```

This shows a perfect linear relationship between the variables, similar to figure (a), with the difference being that the variables are inversely related, i.e., increasing the value of one variable results in a decrease in the other variable.

```
Figure (e)
r = 0.51
```

The relationship between the variables is not very strong, and the data points are a little scattered, although still closer to a straight line.

```
Figure (f)
r = -0.70
```

This is similar to figure (c), with the difference being that the variables are negatively correlated.

We can see that as the value of r decreases, the data points are more scattered, whereas the data points are closer to a straight line when the value of r approaches -1.0 or +1.0.

Q: How is a correlation coefficient interpreted?

A correlation coefficient has both magnitude (unit less) and direction (positive or negative sign) lying between -1 to 1. A correlation coefficient with a value of zero indicates that no relationship exists between the variables. When the correlation coefficient has a non-zero value, and the values approach ±1, this signifies a strong linear relationship between the variables. The magnitude of r signifies the strength of the relationship, and the +ve and -ve signs indicate whether the relationship is directly or inversely related.

For example, if r = 0.9, this means that the variables are strongly related, and increasing the value of one results in an increase in the value of the other. Similarly, if r = -0.9, this indicates that the variables are strongly related, and increasing the value of one results in a decrease in the value of the other.

Q: What are the various issues with correlation?

The various issues with correlation are

- Correlation analysis does not measure the strength of a nonlinear association between variables.

- Accidental or spurious relationships are not accounted for.

- Research problems, such as data contamination, sample bias, etc., hinder drawing reliable conclusions.

- Correlation analysis measures the relationship and does not provide an explanation or basis for it, which can result in false conclusions.

Q: How would you calculate a correlation coefficient in Excel?

A correlation coefficient between two arrays can be calculated in Excel using the CORREL function. The syntax for using the CORREL function is

```
CORREL (array1, array2)
For eg:
CORREL (A1:A20, B1:B20)
```

The CORREL function gives error in the case of the following:

- A #N/A error for unequal numbers of data points in the two arrays

- A #DIV/0! error when the standard deviation of array1 and array2 is zero

Q: What is meant by the term "linear regression"?

Linear regression is a statistical modeling technique that attempts to model the relationship between an explanatory variable and a dependent variable, by fitting the observed data points on a linear equation, e.g., modeling the body mass index (BMI) of individuals by weight.

A linear regression is used if there is a relationship or significant association between the variables. This can be checked by scatter plots. If no association appears between the variables, fitting a linear regression model to the data will not provide a useful model.

A linear regression line takes equations in the following form:

```
Y = a + bX,
Where, X = explanatory variable and
       Y = dependent variable.
       b = slope of the line
       a = intercept (the value of y when x = 0).
```

Q: What are the various assumptions that an analyst takes into account while running a regression analysis?

Regression analysis depends on the following assumptions:

- The relationship between the variables should be linear (or approximately linear) over the range of population being studied.

- All the variables in the regression analysis should be normal, i.e., should follow the normal curve (exactly or approximately).

- There should be no multicollinearity, i.e., the independent variables should not show correlation among themselves.

- There should be no autocorrelation in the data, i.e., the residuals should be independent of each other.

- The condition of homoscedasticity, i.e., the error terms or residuals along the regression, should be equal.

Q: How would you execute regression on Excel?

Regression on Excel can be performed by using three built-in functions to calculate slope, intercept, and R^2 values or by using the Regression function provided in the Data Analysis toolbar (after installing Analysis ToolPak add-ins). The built-in functions are SLOPE(), INTERCEPT(), and RSQ().

Q: What is the multiple coefficient of determination or R-squared?

The multiple coefficient of determination, R^2, is a method by which to calculate the overall effectiveness (in terms of percentage similar to linear regression) of all the independent variables in explaining the dependent variable.

For example, if $R^2 = 0.8$, this means that the independent variables have 80% of the variation in the value of dependent variables.

Unfortunately, R^2 alone may not be a reliable measure of the accuracy of the multiple regression model, as R^2 increases every time a new variable is added in the model, even though the variable might not be statistically significant. If there is a large number of independent variables, the value of R^2 may be high, even though the variables do not explain the dependent variable that well. This problem is called overestimating the regression.

By adjusting the R^2 value for the number of independent variables, the problem of overestimating the regression can be overcome.

Q: What is meant by "heteroscedasticity"?

When the variance of the residuals differs across observations in the sample, this is called heteroscedasticity. It is one of the errors in regression analysis that analysts have to test before running the regression analysis. One of the assumptions of multiple regression is that the variance of the residuals is constant across observations.

Q: How do you differentiate between conditional and unconditional heteroscedasticity?

Unconditional heteroscedasticity occurs in cases in which the level of independent variables does not affect heteroscedasticity, i.e., it doesn't change systematically with changes in the value of independent variables. Although this is a defilement of the equal variance assumption, it frequently causes no serious problems with the regression.

Conditional heteroscedasticity is heteroscedasticity that is related to the level of (i.e., conditional upon) the independent variables.

Q: What are the different methods of detecting heteroscedasticity?

There are two methods of detecting heteroscedasticity: examining scatter plots of the residuals, and using the Breusch-Pagan chi-square test. Plotting the residuals against one or more of the independent variables can help us spot trends among the observations (see Figure 5-2).

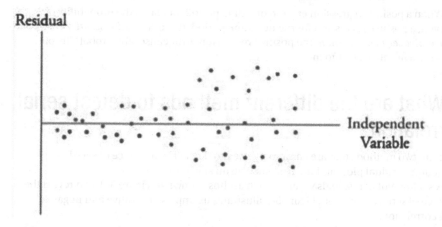

Figure 5-2. *Plotting residuals against independent variable*

The residual plot in the figure indicates the presence of conditional heteroscedasticity. Notice how the variation in the regression residuals increases as the independent variable increases. This indicates that the variance of the dependent variable about the mean is related to the level of the independent variable.

The more common way to detect conditional heteroscedasticity is the Breusch-Pagan test, which calls for the regression of the squared residuals on the independent variables. Independent variables contribute significantly in explaining squared residuals in case of conditional heteroscedasticity.

Q: What are the different methods to correct heteroscedasticity?

The most common remedy is to calculate robust standard errors. The t-statistics is recalculated using the original regression coefficients and the robust standard errors. A second method to correct for heteroscedasticity is to use generalized least squares, by modifying the original equation.

Q: What is meant by the term "serial correlation"?

Serial correlation, or autocorrelation, is the phenomenon commonly observed in time series data, in which there is a correlation between the residual terms. It is of two types: positive and negative.

When a positive regression error in one time period increases the probability of observing a positive regression for the next time period, this is a *positive serial correlation*. In a *negative serial correlation*, the positive regression error causes the probability of observing a negative error to increase.

Q: What are the different methods to detect serial correlation?

There are two methods that are commonly used to detect the presence of serial correlation: residual plots and the Durbin-Watson statistic.

A scatter plot of residuals vs. time, such as those shown in Figure 5-3, can reveal the presence of serial correlation. Figure 5-3 illustrates examples of positive and negative serial correlation.

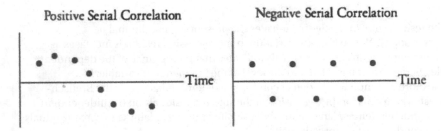

Figure 5-3. *Scatter plot of residuals vs. time indicating positive and negative serial correlations*

The more common method is to use the Durbin-Watson statistic (DW) to detect the presence of serial correlation.

Q: What are the different methods to correct multicollinearity?

The most common method to remove multicollinearity is to omit independent variables having a high correlation with the variable set. Unfortunately, it is not always an easy task to identify the variable(s) that are the source of the multicollinearity. There are statistical procedures that may help in this effort, such as stepwise regression, which systematically removes variables until multicollinearity is reduced.

A summary of what you have to know regarding violations of the assumptions of multiple regression is offered in Table 5-1.

Table 5-1. *Summary of Violations of Assumptions of Multiple Regression*

	Conditional Heteroscedasticity	Serial Correlation	Multicollinearity
What is it?	Residual variance related to level of independent variables	Residuals are correlated	High correlation among two or more independent variables
Effect?	Coefficients are consistent. Standard errors are underestimated. Too many Type I errors	Coefficients are consistent. Standard errors are underestimated. Too many Type I errors (positive correlation)	Coefficients are consistent (but unreliable). Standard errors are overestimated. Too many Type II errors
Detection?	Breusch-Pagan chi-square test	Durbin-Watson test	Conflicting t and F statistics; correlations among independent variables if k = 2
Correction?	Use White-corrected standard errors	Use the Hansen method to adjust standard errors.	Drop one of the correlated variables.

Q: What is an odds ratio?

Odds is the relative occurrence of different outcomes, expressed as a ratio of the form a:b. For example, if the odds of an event are said to be 5:2 in favor of the first outcome, this means that the first outcome occurs five times for the second outcome to occur twice. Odds are related to probability and can be shown mathematically as follows:

```
Odds = a:b
Probability = a/(a+b)
Probability = Odds/(1 + Odds)
Odds = Probability/(1 - Probability )
```

Q: How is linear regression different from logistic regression?

Linear regression is applicable on numerical or continuous variables, but logistic regression is applicable when the dependent variable is categorical (a commonly dichotomous variable). The output of logistic regression is between 0 and 1, where 1 denotes "success" and 0 denotes "failure." But in linear regression, the output is continuous, which can assume any range of value.

Linear regression predicts numerical outputs, such as sales or profit, whereas logistic regression predicts dichotomous output, such as yes and no or living and dead.

CHAPTER 6

■ ■ ■

Segmentation

Segmentation is the process of dividing a data set into clearly differentiated groups, relevant to a particular business. Companies segment their customers into different clusters to decide how to create differentiated strategies for each cluster and to maximize the value of the business.

But segmentation is not just used for customer analytics; it can be used for myriad other solutions. We might want to segment geographical areas based on their population density, or employees on their propensity to attrite.

Segmentation algorithms (also known as clustering algorithms) are very common, and the chances are extremely high that you will be tested on these concepts in an interview. In this chapter, I will go through some common interview questions related to segmentation and clustering.

Q: What are supervised and unsupervised learning algorithms? How are they different from each other?

Supervised and unsupervised learning algorithms are the two broad classifications for all statistical algorithms. The major difference between the two is how outputs to a model are defined. Keeping this segregation in mind helps an analyst to better choose which kind of problem-solving is best suited to a situation.

In supervised learning, we have an heuristic output already present, and the model defines the cause and effect of inputs on given outputs. In other words, the inputs define what we are looking for in a model. So, in supervised learning models, we focus the model on existing relationships between inputs and outputs, to define and forecast the unknown. For example, in all classification techniques, we know from historic data what the different categories in dependent variables are. The goal of these techniques is not to come up with categories but to define them and how they are dependent on independent variables.

In unsupervised learning, the output of the model is undefined. The unsupervised learning models are used to define what our output looks like. Clustering techniques are a classic example of unsupervised learning. Here, we do not have the clusters (output) beforehand; rather, the model comes out with the clusters based on the input and criteria.

B. Gupta, *Interview Questions in Business Analytics*, DOI 10.1007/978-1-4842-0599-0_6

Q: Can you give an example to differentiate between supervised and unsupervised learning algorithms?

The example I'll use here is face recognition.

- *Supervised learning*: Learning, from examples, what a face is, in terms of structure, color, etc., so that after several iterations, the algorithm can define a face.

- *Unsupervised learning*: Because the example provided does not yield a desired output, categorization is undertaken, so that the algorithm differentiates correctly between the face of a horse, cat, or human (clustering of data).

Q: What are some of the supervised and unsupervised algorithms?

All classification algorithms fall under the supervised category. Following is a list of a few classification techniques:

- Naïve Bayes

- Support vector machine

- Random forest

- Decision tree

- Logistic regression

All segmentation algorithms and variable reduction algorithms (such as those in the following list) fall under the unsupervised category.

- K-means

- Fuzzy clustering

- Hierarchical clustering

- Factor analysis

Q: How is clustering defined?

Clustering (or segmentation) is a kind of unsupervised learning algorithm in which a data set is grouped into unique, differentiated clusters.

Let's say we have customer data spanning 1,000 rows. Using clustering, we can group the customers into differentiated clusters or segments, based on the variables. In the case of customers' data, the variables can be demographic information or purchasing behavior.

Clustering is an unsupervised learning algorithm, because the output is unknown to the analyst. We do not train the algorithm on any past input-output information, but let the algorithm define the output for us. Therefore (just like any other modeling exercise), there is no right solution to a clustering algorithm; rather, the best solution is based on business usability.

Q: What are the two basic types of clustering methods?

There are two basic types of clustering techniques:

- Hierarchical clustering
- Partitional clustering

Q: How is hierarchical clustering defined?

Hierarchical clustering attempts to either merge smaller clusters into larger ones or break larger clusters into smaller ones. The basic rule at the core of this technique is deciding how two small clusters are merged or which large cluster is split. The final outcome of the algorithm is a tree of clusters called a dendrogram, which displays how the clusters are related. By splitting the dendrogram at a chosen level, a clustering of the data set into separate groups is achieved.

As shown in Figure 6-1, the dendrogram is split at various levels to come up with the required number of clusters.

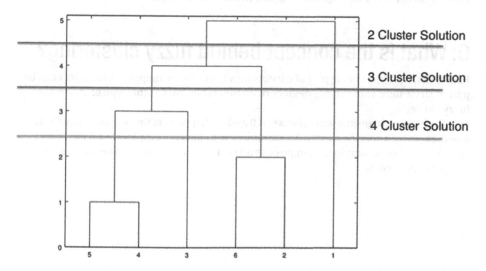

Figure 6-1. *Dendrogram and cluster solutions*

Q: How is partitional clustering defined?

Partitional clustering works to directly decompose the data set into a set of differentiated clusters. The core rule here is to minimize some measure of dissimilarity in the samples within each cluster, while maximizing the dissimilarity of different clusters.

For example: At most times, researchers try to reduce the within-cluster variance and increase variance between the cluster. A good measure is to take the ratio of the 2 measure and maximize it to ascertain the right number of clusters.

Q: What is meant by "exclusive clustering"?

This is the most common type of clustering, in which each object or data point belongs exclusively to only one cluster. This is also the most desired form of clustering, in most cases. For example, it would be necessary for a customer to be part of only one segmentation group, so that a unique, dedicated, and exclusive marketing effort could be formulated as part of a campaign.

Q: What is non-exclusive or overlapping clustering?

Often, if not always, an object can be part of more than one cluster. These are mostly borderline objects, in which we define the boundaries of clusters to overlap each other.

An example is demographic clustering, in which students can be part of both a student cluster and a high-spender cluster, which would be rare.

Q: What is the concept behind fuzzy clustering?

Rather than an object being part of clusters only (one-to-one mapping), an object can be part of all clusters, with varying degrees of membership. We call this type of clustering fuzzy clustering.

Each object is given a score (between 0 and 1) that depicts the degree to which an object is part of a specific cluster. An example is the cluster of employees on the basis of their skill sets. An employee can possess all skill sets but exhibit varying degrees of competency in each.

Q: Can you differentiate between a complete vs. a partial clustering?

In a complete clustering, all objects in the data sets are forced to be part of a cluster. Even when there are outliers in the data sets, they are definitely attached to a cluster or are clusters in themselves.

On the other hand, in a partial cluster, all data points are not necessarily part of a cluster. An example of this can be employees with different skill sets. An employee can be part of more than one cluster of skills.

Q: What is meant by "k-means clustering"?

K-means is the most widely used clustering algorithm in industry. The reason for its popularity is based on the fact that it's both easy to execute and understand.

At the crux of it, k-means clustering identifies random means centers in data sets and attributes cluster membership around those means.

As shown in Figure 6-2, random points are determined, which later form the cluster centers.

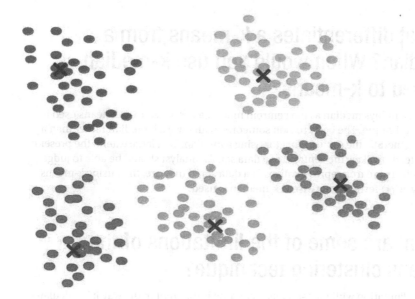

Figure 6-2. *K-means clustering*

Q: What is the basic algorithm of k-means clustering, in layperson's terms?

We first choose the k random means from our data sets. K is the number of clusters that we would like to finally extract from our data set.

Now, each data point is attached to each mean that we have chosen, based on the proximity of that data point to the mean. The group of all these data points with their respective mean will form a cluster.

We then recompute the mean for each computed cluster. The preceding steps are rerun until the recomputed means of all clusters are correctly determined.

Q: What is the proximity measure that you take in k-means clustering?

There are various proximity measures that can be employed. *Euclidean distance* is one such measure that is heavily used. It is simply the ordinary distance between two points. Other measures include Taxicab metric, Manhattan distance, and Jaccard measure.

Q: What differentiates a k-means from a k-median? When would you use k-median as opposed to k-means?

A k-median employs median as the centroid metric, as opposed to the means used in a k-means technique. The basic reason someone would use a k-median rather than a k-means is generally the same as that for using a median. To a large extent, the presence of outliers tends to skew the centroid of a data set. An analyst should be able to judge whether outliers are true representatives of a data set. If they are, then using k-means would may be preferred; otherwise, a k-median is used.

Q: What are some of the limitations of the k-means clustering technique?

The biggest limitation with the k-means technique is inherent in the way it is calculated. The user is required to know beforehand the number of clusters that he or she intends to extract from the data set. This can be both a positive or a negative. It can be positive, because the algorithm is forced to give out the number of clusters that the user requires for business execution, irrespective of whether there is a better cluster solution.

On the other hand, it is a limitation, because the user is not informed whether there is a better cluster solution for the data set. A seven-cluster solution might be better than a four-cluster solution. Analysts usually run all cluster solutions and then pick the one that is most efficient or makes most business sense.

The other biggest issue with the k-means technique is the fact that the algorithm can give different results in different iterations. This is because of the way this technique is designed. Because the first step involves identifying random centroid values, each iteration would have different values and thus can give different results.

Q: Given that each iteration of k-means gives us different results, how would you ensure picking the best results?

This is done by calculating the SSE for each iteration. SSE stands for *sum of squared error*. SSE is calculated by first determining the distance between each data point and closest computed centroid and then summing all these distances. A smaller SSE represents a better solution.

Q: What are the two types of hierarchical clustering?

The two types are

- Agglomerative clustering
- Divisive clustering

Q: What is the difference between agglomerative vs. divisive clustering?

The main difference lies in how the initial group is defined. In agglomerative clustering, each data point is considered a cluster of its own. In each iteration, the data points are merged to form clusters that eventually form one big cluster containing all data points. Consider this a bottom-up approach.

On the other hand, divisive clustering is a top-down approach, in which all data points are initially considered part of one big cluster and then eventually broken into sub-clusters. Finally, the optimal number of clusters is derived.

Q: What is a dendrogram?

A dendrogram is a graphical representation (as illustrated in Figure 6-3) of data sets and their cluster membership, using a tree-like diagram. Each vertical line represents either a data point or a cluster. The bottom-most vertical lines represent a data point. As we subsequently move up the diagram, the vertical lines merge (finally, into one), to reveal the cluster formation.

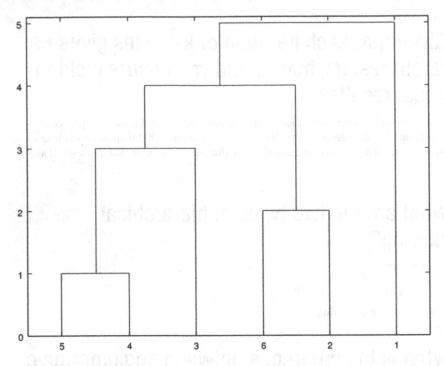

Figure 6-3. *A dendrogram*

Q: What is the basic algorithm behind the agglomerative clustering technique?

First, consider all data points as a separate cluster. Then, using a proximity measure, define the proximity between all clusters. Combine clusters that are closest. Repeat this until only one cluster is left.

Q: Can you briefly explain some of the proximity measures that are used in hierarchical clustering techniques?

There are numerous proximity measures used in the clustering techniques. *Min* defines cluster proximity as the distance between the closest two points in the clusters, whereas *max* defines cluster proximity as the maximum distance between any two points.

Group average refers to the average of all pair-wise distance between the points in the clusters.

An alternative technique, *Ward's method*, is more widely used.

Q: What is Ward's method of defining cluster proximity?

Ward's method is very similar to the k-means method of finding optimal cluster numbers. It measures the proximity by the increase in SSE when two clusters are merged. In other words, it reduces the sum of squared errors while clusters are joined together.

Q: How do you determine the optimal number of clusters for a data set?

A clustering technique is both an art and a science. Determining the optimal number of clusters is a crucial part of a clustering technique, and it is different from the actual clustering itself.

Determining optimal clusters requires consideration of both the technical aspects as well as the business aspects. One such technical methodology is the *elbow method*.

Q: Can you briefly explain the elbow method to determine the optimal cluster solution?

An elbow plot (Figure 6-4) is a graph drawn from a number of clusters on an x axis and the SSE of each cluster solution on a y axis. We also call this a hockey stick graph, because the curve bends sharply, like an elbow point. We consider the point on the bend of the x axis as the optimal cluster solution.

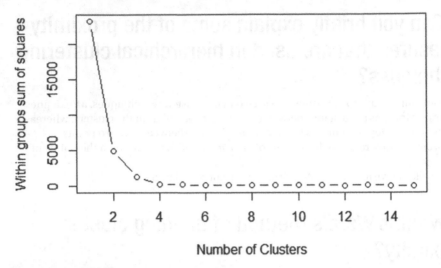

Figure 6-4. *Elbow plot*

In the following figure, we notice that the curve bends sharply at cluster number 4. The incremental reduction in error terms, while increasing the number of cluster solutions after 4, is very low. Thus, we consider 4 as the optimal cluster solution.

Q: What is the business aspect of determining the optimal cluster solution?

A clustering algorithm only segments a data set into various clusters, in which data sets are closer together, based on various attributes. Defining these clusters into meaningful definitions and identifying usage and business strategy around them is something that an analytical person or data scientist brings to the mix.

Each cluster identified by an algorithm should have a business meaning. For example, demographic segmentations can yield clusters that are meaningless for business usage. Also, clustering solutions cannot identify segments that actually have some meaning. For example, attributes with absolute zero values cannot be identified by the algorithm, as they almost always cluster data sets in a vicinity.

This is where the business aspect of the solution comes into the picture. Most of time, analysts come out with a high number of optimal cluster solutions, using technical aspects. Then they merge various clusters, using business rules. As stated previously, this is as much an art as a science.

Q: Can you explain, using a case study, the use of clustering techniques in the retail industry?

A mobile phone manufacturer would like to launch in a new geographical area. Traditionally, the phone manufacturer competes at all price levels and custom-creates phones to suit different buyers in a particular area.

For the new geographical area, the manufacturer starts out by clustering the large sample data set of citizen demographics. This data set is then enriched using a primary survey of needs of various mobile phone users.

Using this clustering methodology, the manufacturer is able to distinctly identify five clusters that behave in different ways and have unique needs. Thus, the manufacturer custom-creates a mobile phone for these five segments. Subsequently, marketing campaigns are also designed by keeping the behavior of these five segments in mind.

CHAPTER 7

■ ■ ■

Advanced Statistics and Usage

I have covered some basic concepts and techniques of statistics in earlier chapters. Although statistics itself is an extremely broad discipline, the goal of this book is to highlight concepts that prospective candidates for jobs in the field of analytics will most likely be tested on during an interview.

In this chapter, I will briefly cover various advanced topics in statistics, in addition to their various industrial applications.

One of the common questions interviewers ask to test the knowledge of a prospective data scientist is to discuss some use cases in various business areas. I will try to cover as many such cases as possible here.

Q: What is understood by one-way analysis of variance?

The one-way analysis of variance (ANOVA) test is used to determine whether the mean of more than two groups of a data set are significantly different from each other.

Imagine, for example, that we are conducting a BOGO (buy one get one) campaign involving five groups of a hundred customers each. Each group is different in terms of its demographic attributes. We would like to determine whether these five groups respond differently to the campaign. This would help us to optimize the right campaign for the right demographic group, increase the response rate, and reduce the cost of campaign.

Q: In a nutshell, how does the ANOVA technique work?

The "analysis of variance" works by comparing the variance between the groups to that within the group variance. The core of this technique lies in assessing whether all the groups are, in fact, part of one larger population or a completely different population with different characteristics.

© Bhasker Gupta 2016

B. Gupta, *Interview Questions in Business Analytics*, DOI 10.1007/978-1-4842-0599-0_7

Q: What is the null hypothesis that ANOVA tests?

The null hypothesis is

$$H_o = \mu_o = \mu_1 = \mu_2 = \mu_3 = \ldots\ldots = \mu_k$$

where μ = the group mean and $_k$ = number of groups.

In a null hypothesis, the means of all groups are equal to one another. If there are even two groups with significantly different means, then we accept an alternate hypothesis.

Q: What is understood by dimension reduction techniques?

Dimension (variable) reduction techniques aim to reduce the data set with higher dimension to one of lower dimension, without the loss of feature of information that is conveyed by the data set. The dimension here can be conceived as the number of variables that a data set contains.

With the advent of big data and the ability to process and store large amounts of data, organizations today try to store as much data as possible. This leads to an increase in the attributes that are stored. Think of it as a database table that increases not just in rows but also in terms of columns (variables).

For data scientists creating appropriate models, not all variables are relevant. In addition, large multicollinearity, which diminishes model performance, may be encountered. Variable reduction techniques help weed out this issue. Also, the model is much more crisp in terms of being able to be understood and explained and is less costly (uses less computational resources).

Dimension reduction techniques are almost always executed as a precursor to another technique, such as regression. It is a way to speed up model-building without compromising on the potential of a model.

Q: What are some commonly used variable reduction techniques?

Two commonly used variable reduction techniques are:

- Principal component analysis (PCA)
- Factor analysis

Q: Can you provide a brief overview of principal component analysis?

The crux of PCA lies in measuring the data from the perspective of a principal component. A principal component of a data set is the direction with largest variance. A PCA analysis involves rotating the axis of each variable to the highest eigenvector/eigenvalue pair and defining the principal components, i.e., the highest variance axis or, in other words, the direction that most defines the data. Principal components are uncorrelated and orthogonal.

Q: Can you provide a brief overview of factor analysis?

The key concept behind factor analysis is the presence of a latent variable that stores much of the information of a set of variables in a data set. For example, a group of respondents can answer questions relating to income, education, and spending similarly, because they are in the same socioeconomic category.

In factor analysis, we define factors that are the same in number as the number of variables in a data set. Each factor captures a certain amount of variance in each variable. The *eigenvalue* is the measure of how much variance of observed variables is captured by a factor.

All factors are sorted in their descending order of value. The factors with low value are discarded, and top factors are retained as factors that explain most variance in the observed variance. It is helpful to know the number of factors in advance.

Q: What is factor loading?

Each factor in a data set defines the latent variable, in other words, the underlying variable that defines a set of variables in a data set. Factor loading describes the relationship or association between each variable and each factor. Higher association indicates that the factor can be used to describe that variable.

An example: While analyzing 70 variables that affect customer churn, a factor analysis was run, and because 70 variables were used for this analysis, the algorithm gave 100 factors. On observing the factor loading, it is found that demographic variables have high loading for one specific factor. So, these variables are combined into one factor, and so on, for other variables.

Factor loading is an important parameter to assess factor-variable dependence.

Q: What is conjoint analysis?

Conjoint analysis is widely used in market research to identify customers' preferences for various attributes that make up a product. The attributes can be various features, such as size, color, usability, price, etc.

Using conjoint (trade-off) analysis, brand managers can identify which features customers would trade for a certain price point. Thus, it is a much-used technique in new product design or for formulating pricing strategies.

Customers undergo a carefully designed survey, which showcases products with different attributes, and customers are then asked the likelihood of their purchasing the products. As shown in Figure 7-1, rather than asking their preference up front, the survey showcases different product-price combinations, to determine the latent needs of customers in context with the right price point.

Which of the following men's face wash would you choose?

	Brand 1	Brand 2	Brand 3
Price	$4.99	$7.99	$6.99
Type of Top	Flip top	Screw top	pull top
Viscosity	High	Medium	Low

Figure 7-1. A typical survey designed for conjoint analysis

Q: What are the three main steps involved in executing a conjoint analysis?

First, the product has to be divided into attributes and features.

The second step involves how these attributes have to be presented to the survey respondents. This step also includes deciding which rating methodology should be used for these attributes to be picked by respondents. The design decision takes into account factors such as number of respondents, time available for each response, complexity of product, and its features.

The third step is the statistical algorithm itself. The part-worth model is one of the simplest models available to assess the utility of each attribute.

Q: What are some of the ways in which an HR department can use analytics?

Human resource (HR) analytics forms the crux of the human resource (HR) function in most organizations and is a complex application of data mining and statistical techniques applied to employee-related data. HR analytics is a crucial technique for quantitatively measuring the outcome of employee engagement programs, and it provides corporations with the power to measure year-on-year contrasts with regard to numerous factors.

HR analytics provides organizations with powerful insights to help efficiently manage employees to increase productivity. There are various challenges in this field, including identifying the right data, capturing, storing and processing that data, and building the right model to increase return on investment (ROI) of the analytics function.

HR analytics finds its application in various business functions. The core functions of HR, such as recruitment and training, mergers and acquisitions, designing compensation structure, and improving performance appraisal processes are revolutionized by applying analytics to the historical data.

Some emerging niche areas in HR analytics are

- Employee sentiment analysis
- Predictive attrition models
- Net promoter score

Q: What are some of the specific questions that HR analytics helps to answer?

Some questions that HR analytics helps to find answers to include the following:

- Identifies the motive behind high employee attrition in an organization
- Identifies the reasons for lagging performance
- Finds strategies to improve team efficiency
- Measures the gaps in employee skills and identifies ways to fill them
- Identifies efficiencies in employee orientation (or comparable policies)
- Forecasts who is right to assume a specific role
- Makes the correlation between performance rating and employee performance

Q: What, briefly, is employee sentiment analysis?

Sentiment analysis involves extracting meaning and insights from large amounts of structured and unstructured data available to HR via various sources. These can be both internal and external sources. Internal sources include annual surveys, internal blogs and e-mails, etc. External sources include social media channels, blogs, and external surveys.

The data should be continuously tracked, analyzed, and scrutinized on key topics, which can be used to understand how employees feel about the company and what can be changed to make it a better place to work. Glassdoor.com, LinkedIn, etc., are employee-related portals continuously being used by various companies, to run sentiment analysis.

Q: What, in detail, is the predictive attrition model?

Employee attrition is predictable under stable circumstances, wherein a set pattern can be deduced from certain parameters influencing the employee and the organization at all times. Some of these parameters could be foreseeable, such as retirement age, or unforeseeable, such as company performance, external funding, management shake-up, etc.

However, who is going to leave, when, and why can be answered, based on analytical models developed as a result of data analysis.

Through predictive algorithms, companies gain better understanding and can undertake preventive measures to counter employee attrition.

On a basic level, the model works by clustering/classifying employee profiles, based on various attributes, such as age, sex, marital status, education level, work experience, distance from hometown, etc., and generates various levels of risk of attrition. Occasionally, other parameters, such as performance over the years, pay raise, work batch, and educational institution, are also taken into consideration.

However, the accuracy of the model is directly proportional to the selection of parameters, which, in turn, leads to the generation of the "type" of predictive model most suitable for the organization.

A predictive attrition model helps not only in taking preventive measures but also in making better hiring decisions. Deriving trends in a candidate's performance from past data is important for predicting future trends, as well as to gauge new employees. Moreover, HR can use the employee data to predict attrition and the possible reasons behind it, and can take appropriate measures to prevent it.

We live in a data-driven world. From something as trivial as weather updates to GPS navigation, data is constantly being generated every second, in every field, which leaves us to decide how to turn it around to our advantage in our respective domains.

Q: How would you create a predictive attrition model? What, briefly, is the statistics part of this model?

Various statistical and machine-learning algorithms are designed to construct predictive models. For instance, *classification* models catalog employees based on their risk of leaving a company, whereas *nonlinear regression* models determine the *probability of attrition* when outcomes are dichotomous.

Likewise, a *decision tree* model evaluates loss based on such factors as Gini index, information gain, and variation reduction. For models involving multiple parameters, decision trees tend to become very large and complex.

In such circumstances, a *random forest* method combines several decision trees, using multiple algorithms to classify and understand complexities and predictions. In addition, these models aim to provide good predictability. However, seamless implementation depends on choosing the right model.

Thus, different models are chosen, based on the aforementioned parameters, data availability, budget, computational power, and the requirements of decision makers. For example, in one organization, a model using an artificial neural network may provide better predictability than a decision tree model, but a decision tree model may be easier to understand and implement at a lower cost.

Thus, depending on the organizational contexts, different models have to be tried and evaluated before making the final selection.

Q: What would a typical output of a predictive attrition model look like?

The output depends on the chosen model. For instance, a *logistic model* produces scorecards for employees, based on their predicted *attrition risk* parameters, while the classification model catalogs employees into wider parameters, such as more likely or less likely to quit, high risk or low risk, etc. Figure 7-2 gives the output of a typical logistic regression model, in which employees are scored on a retention predictor and charted on a distribution graph.

However, the bottom line is to keep it simple enough for HR managers to understand and implement. Changing the various factors helps in assessing the impact of changes and making the right decisions.

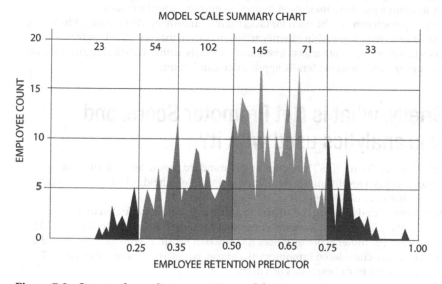

Figure 7-2. *Output of a predictive attrition model*

Q: What are some of the benefits of a predictive attrition model?

This model is helpful in doing the following:

- Evaluating employee requirements, strengths, and weaknesses
- Minimizing the cost of new talent acquisition, based on employee profiling and company requirements
- Analyzing and assessing loss in expertise and skill sets
- Measuring financial and productivity loss due to attrition
- Enabling planning and minimizing loss
- Providing a good understanding of workforce supply and demand
- Enabling the preparation of contingency plans, based on the insight and foresight provided by the model

Q: How can HR managers use analytics to promote employee effectiveness?

We can use a variable reduction technique such as factor analysis to identify the right attributes that will have the most impact on employee productivity. So, rather than focusing on many aspects of employee engagement, we can run statistical algorithms to identify attributes that have maximum impact on employee performance.

This approach can also be used for tasks such as improving skills, leadership hiring, employee movement, leadership identification, etc. HR managers can identify the attributes that affect skills at various levels and positions, using statistical algorithms, and can then extrapolate these to identify appropriate candidates.

Q: Briefly, what is Net Promoter Score, and how is analytics used with it?

The Net Promoter Score (NPS') is becoming the standard metric for measuring customer satisfaction. NPS is a loyalty metric developed by Fred Reichheld, a Fellow of Bain & Co. and a board member of Satmetrics.

NPS is calculated by using the answer to a single question, evaluated on a 1–10 scale: "How likely is it that you would recommend [brand] to a friend or colleague?" This is called the Net Promoter Score question. As shown in Figure 7-3, people who award a score of 9 or 10 are considered Promoters of a brand, whereas any score of less than 7 is considered to come from Detractors of a brand.

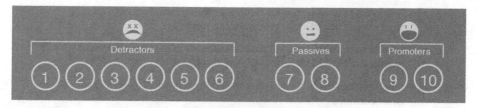

Figure 7-3. Promoters, Passives, and Detractors in the NPS survey

NPS Score = % of Promoters - % of Detractors

In addition to the NPS question, each respondent can rate a brand/company across a number of service-delivery attributes.

Thus, statistical analysis helps to derive that, for each one-point improvement in these attributes, NPS will improve by xx points.

Variables that affect NPS can be assessed by using multiple regression techniques and beta coefficients, according to the following methodology:

1. Relationships are identified by correlating the variables.

2. Stepwise regression evaluates the different beta values for different variables, to identify how much a variable affects NPS.

3. The best beta values, the service-delivery level attributes, and other variables affecting NPS the most are identified.

4. The analysis explains xx% of the variance in NPS, i.e., for each unit improvement in the determined elements, the NPS score is improved by zz points.

5. Those attributes that are most important for all customers are identified at a corporate level, and this provides further scope for improvement.

Q: Briefly, how can analytics be used in the hospitality industry?

In the hospitality industry, analytics can be utilized to expand business operations, optimize marketing strategies, and increase occupancy rates and yield. For example, through analytics

- It allows a concierge to know which local tours are best suited to a guest's preferences, based on his/her past behavior.

- It lets a restaurant forecast the menu entries that are most expected to be ordered, based on what the weather is expected to be for the day.

- It helps in pricing decisions, such as, given the occupancy rate, what should be the right pricing of the room(s).

- It helps to optimize marketing strategies, such as what offers and campaigns should be sent to whom and when.

- It helps hoteliers cut down their energy costs without sacrificing guest comfort.

Q: Can you describe a use case of social media analytics

Social media has recently emerged as a goldmine of consumer-related information that supplements traditional ways of collecting information. Customers freely talk about their preferences and what they are interested in.

This data may include affinity toward a genre of music or a specific news website. Social affinity data can be a powerful way to help understand how brands relate to people's many interests, such as musicians, books, websites, movies, or celebrities.

This information is highly useful for brands, as it unearths deep-rooted preferences among customers, which might otherwise be difficult to identify.

By mining thousands of conversations and engagements over social media channels, marketers can identify key affinities for their brands. This can also be in the form of interest segments that align with your brand or with your competitors'.

Q: What is understood by text analytics?

Text analytics is the process of deriving high-quality information from free-flowing text. Organizations today have much more unstructured text than they realize. This may be in the form of comments from sales teams and agents, minutes of meetings, internal chats, e-mails, blogs, and much more. This text stores a vast amount of valuable information that most of the time goes unnoticed. Figure 7-4 details the text analytics inputs such as public and web text, as well as private text, such as a company's internal data.

Using advanced natural language processing tools, we can derive insights from the text. For example, companies can mine thousands of tweets, to understand the customer sentiment for a brand.

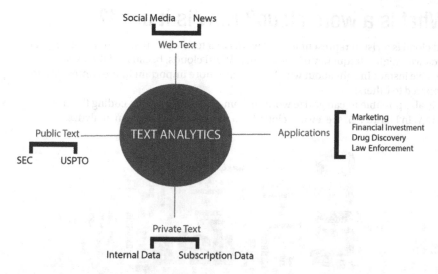

Figure 7-4. Typical text analytics process

Q: What typical process would you undertake for text analytics?

Text analytics involves a three-step process:

1. *Text mining*: More often than not, text might not be as easily
    available as imagined. Large amounts of unstructured data
    reside on various portals, blogs, handwritten material, etc.
    Text mining refers to the process of gathering this data, using
    algorithmic means.

2. *Text parsing*: Text parsing refers to the process of converting
    unstructured text into a form that is more readily analyzable.
    This may include writing algorithms simply to cherry-pick
    information from a text. Or, alternatively, to clean up the text
    (removing punctuation, numbers, stop words, or parts of
    speech tagging), to polish it into a well-rounded, analyzable
    raw data.

3. *Text analysis*: This is where the crux of the whole process
    resides—extracting those interesting nuggets that
    are insightful for the business. Techniques in natural
    language processing comes in handy, including pattern
    recognition, word frequency distribution, sentiment analysis
    (polarity), latent semantic analysis, word classification
    and categorization, single value decomposition, and word
    correlation.

Q: What is a word cloud? How is it used?

A word cloud is a visual representation of words in a text, with greater prominence given to words with higher frequency of appearance. Word clouds, because of their visual nature, give instant insight about which words are more important in the context of a text, as compared to others.

It is also possible to categorize words in some manner by color-coding them separately. In Figure 7-5, see a word cloud from the Wikipedia page on analytics.

Figure 7-5. *Word cloud*

Q: Briefly, how can analytics be used in the banking industry?

The banking, financial services, and insurance (BFSI) industry was an early adaptor of analytics. Today, almost all bank and financial institutions use analytics aggressively, and the industry is full of some very meaningful use cases that have helped it hugely.

Analytics tools give banks insights into the personal habits of its customers, allowing them to promote offers accordingly. Analytics is also used by banks to reduce the chances of money laundering, by identifying suspicious activity, such as moving money to multiple accounts, finding large single-day cash deposits, the opening of a number of accounts in a short period of time, or sudden activity in long-dormant accounts.

Using analytics, a bank is also able to keep track of the credit histories of customers and can distribute loans accordingly.

When salespeople are pitching a loan to a client, a bank tries to find out the background of the customer and what the likelihood of his/her taking a particular loan is. Banks also use analytics to increase customer loyalty and reduce loan prepayments due to refinancing with other institutions.

To follow the AML (anti money-laundering) guidelines around fraud analysis, banks deploy tools that can identify complex schemes/transactions.

The tool provides link analysis to investigate financial similarities between apparently unrelated accounts, to detect money being moved across accounts. It helps in establishing connecting patterns in potentially fraudulent transactions, by scanning a history of transactional data.

Banks are also applying their data models to education loans, automotive loans, housing loans, and loans to small- and medium-sized enterprises (SMEs), to try and reduce the percentage of those loans going bad.

For instance, in the case of an education loan, banks combine data from their bad loans, income tax departments, and credit ratings agencies to identify suitable candidates and then send them reminder messages on Facebook.

Banks are running studies on which colleges in which cities show the most delinquencies in student loans and how to adjust for the increased risk. They also use analytics to determine where ATM branches should be positioned and how much cash should be placed in them.

Q: Can you provide a use case of analytics in the banking industry?

Customers are increasingly using bank debit cards, and with every swipe, they create critical digital information. Analyzing usage patterns on larger data sets will not only reveal her/his buying preferences, but will also highlight engagements with the bank's affiliated merchants. With these insights, a bank's product-development and partnerships teams are better poised to decide if they should enhance partnerships with existing merchants or go for merchants with new and innovative products. They can also leverage the insights to decide on locations where a particular offer from a merchant may grab more eyeballs and mindshare.

Q: Can you provide some key use cases of geospatial analytics?

Following are three key use cases:

Location Comparison Based on Populated Area Density

Currently, finding the location for your next store, ATM, real estate asset, warehouse, etc., is more or less a market research activity. It's imperative that the right location for your next asset make a huge impact on your top line.

This is where geospatial analytics can help, providing a better estimate of the population around an area. With advanced remote-sensing techniques on geospatial data, we can estimate the population density around a particular location. This, combined with mapping competitors and other important sales generators (establishments that help increasing sales directly/indirectly), can provide additional information about important POIs (points of interest) for the most strategic location.

Strategic Location Identification

Companies are faced with resource crunches when deploying resources at higher concentrations of activity. Essentially, a brand's customer or key activity is spread throughout an area, and it's nearly impossible to cater to all these activities through the limited resources at hand. This leads to a strategic exercise into location identification. This may include identifying location for road shows or deploying personnel who maximize impact to customers or activities.

Geospatial analytics, coupled with data from other sources, can be extremely helpful in identifying these hot spots. The addresses of existing customers can be geocoded on the map, and areas that overlap the most within a one- to two-mile radius of these customers' locations can be identified. To advance it further, historic data such as time and geographic features can be coupled with the preceding and be fed into a machine-learning algorithm, to predict the probability of events.

Area Growth Identification

One key decision parameter for many executives is how a location has changed over a period of time. This may include the number of housing facilities that have come up in the area, the change in cultivated land cover, and new roads and highways. This information can be used to identify areas with growth potential for real estate developers, or for retailers, in deciding where to plan their next store.

High-resolution satellite imagery of land areas can be extracted from different time periods, and the images analyzed using geospatial analytics. This gives the percentage change in land cover or in new housing in the area.

CHAPTER 8

■ ■ ■

Classification Techniques and Analytics Tools

A classification technique plays an important role in the whole of analytics-based decision making. This is probably the most important group of techniques to be learned by an analytics professional. Also, given the breadth and depth of these techniques and their vast usage, you are bound to get many questions on this subject in a job interview. So, arm yourself with some basic concepts.

Further, I will delve into some analytics tools, such as R, SAS, and Tableau, which will surely come up in an interview. Besides, having some basic knowledge of databases, SQL, and big data will help you showcase an all-around knowledge of this subject.

I will also briefly touch upon big data, although it is not within the scope of this book.

Q: What is understood by "classification"?

Classification is the grouping of a data set, based on some predefined criteria. The criteria are usually based on some historic information, and classification tries to classify the data set, based on information received from that historic criteria.

An example: A company wants to have a database of 1 million customers in the United States, including their demographic information. It wants to identify the top 50,000 customers who have highest propensity to respond to an offer campaign.

The company's analyst retrieves past data on response rates for a similar campaign on 200,000 customers. Their response rate is trained on a classification technique that tries to separate respondents with nonrespondents and also create a scorecard for the customers. The model is then executed on a 1-million-customer base, to classify respondents from nonrespondents and pick the top 50,000 respondents who should be sent the new campaign.

Other examples include

- Google identifying whether a mail is spam, based on its content and other information

- Assessing whether an employee would attrite, based on his/her past information

© Bhasker Gupta 2016
B. Gupta, *Interview Questions in Business Analytics*, DOI 10.1007/978-1-4842-0599-0_8

Q: Can you name some popular classification methodologies?

There are numerous classification techniques today. This is probably the most widely studied area and encompasses techniques that are so vast and differentiated from one another that the topic itself is mammoth in proportion.

Some of the more widely known techniques are

- Logistic regression

- Neural network

- Decision tree

- Random forest

- Discriminant analysis

Q: Briefly, what is understood by "logistic regression"?

Logistic regression is the technique of finding relationships between a set of input variables and an output variable (just like any regression), but the output variable, in this case, would be a binary outcome (think of 0/1 or yes/no).

For example: Will there be a traffic jam in a certain location in Bangalore? is a binary variable. The output is a categorical yes or no.

The probability of occurrence of a traffic jam can be dependent on such attributes as weather conditions, day of the week and month, time of day, number of vehicles, etc. Using logistic regression, we can find the best-fitting model that explains the relationship between independent attributes and traffic jam occurrence rates and predict the probability of jam occurrence.

Q: What is understood by "neural network"?

Neural network (also known as artificial neural network) is inspired by the human nervous system: how complex information is absorbed and processed by the system. Just as with humans, neural networks learn by example and are configured to a specific application.

Neural networks are used to find patterns in complex data, and thus can forecast and classify data points. Neural networks are normally organized in layers. Layers are made up of a number of interconnected *nodes*. Patterns are presented to the network via the *input layer*, which communicates to one or more *hidden layers*, in which the actual processing is done. The hidden layers then link to an *output layer*, where the answer is output, as shown in Figure 8-1.

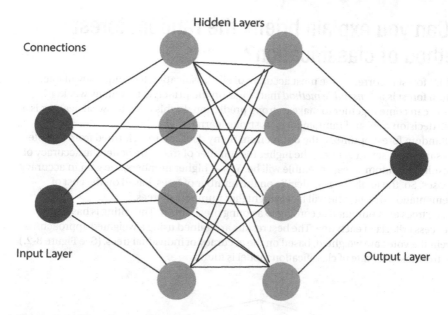

Figure 8-1. *Neural network representation*

Q: How is neural network different from conventional computing?

Conventional computing comprises predefined instructions that form the building blocks of its processing system. Neural networks, on the other hand, do not have predefined steps to processing a system. Rather, they learn from past experiences to chart their own steps in processing.

Q: Can you give a brief overview of decision trees?

Decision trees, as the name suggests, are tree-shaped visual representations by which one can reach a particular decision, by laying down all options and their probability of occurrence. Decision trees are extremely easy to understand and interpret. At each node of the tree, one can interpret what the consequence of selecting that node or option will be.

Q: Can you explain briefly the random forest method of classification?

Random forest is currently the most accurate of all classification techniques available. Random forest is an *ensemble method* that works on the principle that many weak learners can come together to make a strong prediction. In this case, the weak learner is a simple decision tree, and random forest is strong learner.

Random forest optimizes the output from many decision trees formed from samples of the same data set. In general, the higher the number of trees, the better the accuracy of the resulting random forest ensemble will be. Yet, at higher numbers, the gain in accuracy decreases. So, the analyst has to decide on the number of trees, based on the cost of implementation that he/she will face with higher numbers of trees.

The trees are combined according to a voting mechanism. The voting is based on the success criteria of each tree. The best results are gained using a weighted approach, wherein the votes are weighted, based on the accuracy of individual trees. (See Figure 8-2.) Thus, the most accurate of classification model is found.

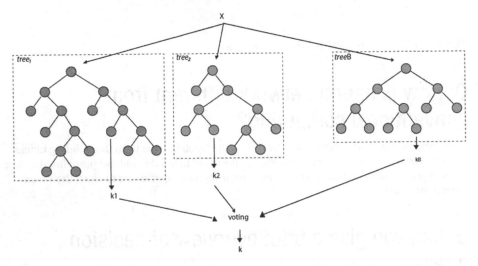

Figure 8-2. *Visual representation of trees in a random forest classification analysis*

Q: Can you explain discriminant analysis, in brief?

Discriminant analysis–based classification works according to the concept of analysis of variance (ANOVA), which is to test whether there is a significant difference between the mean of two or more groups with respect to a particular variable. If the mean of a variable is significantly different in different groups, it can safely be said that this variable classifies the data set into groups.

To extend this concept, MANOVA, or multivariate analysis of variance, can be executed to classify a data set based on multiple variables.

Q: How would you assess the performance of a classification model?

The performance of a classification model is assessed by a table called a *confusion matrix*. It is based on the count of records that are accurately predicted vs. counts of records incorrectly predicted.

Following (Table 8-1) is a confusion matrix for a two-class problem.

Table 8-1. *Confusion Matrix for a Two-Class Problem*

		Predicted values	
		Class 1	Class 2
Actual Values	Class 1	A	B
	Class 2	C	D

Here, A is the number of records of Class 1 that are correctly predicted to be Class 1. B is the count of records of Class 1 that are incorrectly predicted to be Class 2. So, total correct predictions are A+D. Total incorrect predictions are B+C.

Accuracy of a model = Total correct predictions/Total records = A+D/A+B+C+D.

Error Rate of a model = Total incorrect predictions/Total records = B+C/A+B+C+D.

A robust classification model aims to increase the accuracy rate or decrease the error rate of a prediction.

Q: What are some of the visualization tools available in the market today?

Visualization tools can be divided into three broad categories:

Graphical tools:

- *MS Excel*: Microsoft Excel is the standard offering in the Microsoft Office bundle. It is used mostly by analysts for all lightweight analysis, as well as a visualization tool.

- *D3.js*: A JavaScript library to create graphs in HTML and related web technologies

- *FusionCharts*: A JavaScript library for graphs on the Web

- *Google Charts*: Interactive charts for web and mobile devices

Dashboard tools:

- *Tableau*: A US-based software company with a flagship product that helps create dashboards on raw data

- *Qlikview*: A dashboard software prodyct by the US-based company Qlik

- *Spotfire*: Dashboarding software by TIBCO

- *OBIEE*: By Oracle

- *Business Objects*: By SAP

Infographic tools:

- *Infogram*

- *plotly*

- *Picktochart*

Q: How would you define big data?

With both computational power and storage increasing at reduced cost, the amount of data being generated today is profound. Most of that data is either unstructured or semi-structured.

With the advent of this large amount of unstructured data, the traditional method of storing and processing it are inefficient. Big data is large amounts of unstructured data that holds huge potential and value from being mined and stored but, due to its size, could not be stored on traditional database systems.

Q: What are the three *v*'s that define big data?

- *Volume*: The amount of data has to be large, in petabytes, not just gigabytes

- *Velocity*: The data has to be frequent, daily, or even real-time

- *Variety*: The data is typically (but not always) unstructured (like videos, tweets, chats)

Some experts also include a fourth *v*—*veracity*: uncertainty of data.

Q: Can you differentiate between structured, semi-structured, and unstructured data?

Most traditional data is structured, i.e., it can be stored in well-defined rows and columns. Legacy transaction systems are an example of structured data: all transactions are stored in relational database management systems (RDBMSs), with each row representing one transaction and each column representing attributes of that transaction.

Semi-structured data is partially stored in a well-defined database structure. Think of an XML file, which stores data but is not as well-defined as a database table.

Unstructured data cannot be categorized as structured or semi-structured and do not have a well-defined structure associated with how it is stored. Think of most tweets or blogposts. They contain relevant information to be mined, but this information is not structured. Special techniques are applied to extract this information. (See Figure 8-3.)

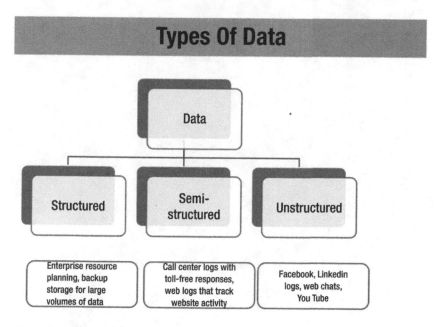

Figure 8-3. *Types of data*

Q: What are some of the tools used for statistical analysis?

Some popular tools for statistical analysis include

- *SAS*: A suite of analytics software developed by SAS, a company based in North Carolina

- *R*: An open source language and environment for statistical computing

- *WEKA*: A suite of machine-learning free software written in Java, developed at the University of Waikato, New Zealand

- *SPSS*: A software for statistical analysis, currently owned by IBM
- *EViews*: Mostly used for econometric analysis, a software developed by Quantitative Micro Software (QMS)
- *Minitab*: A statistical tool developed at Pennsylvania State University

Index

■ T

■ U

■ V

■ W, X, Y, Z

Get the eBook for only $5!

Why limit yourself?

Now you can take the weightless companion with you wherever you go and access your content on your PC, phone, tablet, or reader.

Since you've purchased this print book, we're happy to offer you the eBook in all 3 formats for just $5.

Convenient and fully searchable, the PDF version enables you to easily find and copy code—or perform examples by quickly toggling between instructions and applications. The MOBI format is ideal for your Kindle, while the ePUB can be utilized on a variety of mobile devices.

To learn more, go to www.apress.com/companion or contact support@apress.com.

Printed in the United States
By Bookmasters